Rural Women in the Soviet Union and Post-Soviet Russia

This is the first full-length history of Russian peasant women in the twentieth century in English; it fills a significant gap in the existing literature on rural and gender studies of twentieth-century Russia and is the first to take the story into the twenty-first century. The book offers a comprehensive overview of regulations concerning rural women: their employment patterns; marriages, divorces and family life; and issues with health and raising children. Rural lives in the Soviet Union were often dramatically different from the common narrative of Soviet history. The lives of rural women were even more demanding than those of other Soviet women, and even during the Khrushchev 'Thaw' in the late 1950s and early 1960s, rural women were excluded from the reforms and liberating policies it promoted.

The author, Luibov Denisova, is a leading expert in the field of rural gender history in Russia. She includes material from previously unavailable or unpublished collections and archives in Moscow, St Petersburg, Archangel, and Vologda, as well as interviews and sociological research conducted in 30 different Russian villages, alongside oral traditions such as folk songs and *chastooshkas* among peasant women in Russia. Overall, the book is a history of all rural women, from ordinary farm girls to agrarian professionals to prostitutes; it paints a unique and complete picture of rural women's life in the Soviet Union and post-Soviet Russia.

Liubov Denisova is Professor of History at the Russian State University of Oil and Gas, and is the leading expert in the field of rural and gender history in Russia. Her books include the bestselling *Zhenshchiny russkikh selenii* (Women of Russian Villages) and *Sud'ba russkoi krestianki* (The Fate of Russian Peasant Women).

Irina Mukhina is Assistant Professor of History at Assumption College, Massachusetts, USA. She is a bilingual expert on gender issues during and after Stalinist ethnic deportations, and the author of *The Germans of the Soviet Union* (also published by Routledge).

"The research presented here for the first time in English, is a pioneering work on peasant women and the family. Brilliantly researched, this book will revise our understanding of life in the Russian village in the 20th century."

—Professor Lynne Viola,
University of Toronto, Canada

"This epic book about Russian village and the great role that rural women played in it will undoubtedly open a new era in the existing historiography and will become a new voice in telling the history of Russia. It will resonate in hearts and minds of readers inside and outside Russia."

—Professor Nikolai Ivnitskii,
Russian Academy of Sciences

"This volume explores a topic rarely addressed by historians: the private lives of Soviet women in the countryside. This is an important book that will stimulate debate on women's position under both the Soviet and post-Soviet regimes, and will underscore the need for contemporary Russian sociologists, feminists, and policymakers to distinguish between urban and rural women in their discussions of women and women's issues."

—Professor Roberta Manning, Department of History,
Boston College, USA

Rural Women in the Soviet Union and Post-Soviet Russia

Liubov Denisova
Edited and translated by
Irina Mukhina

Routledge
Taylor & Francis Group

LONDON AND NEW YORK

First published 2010
by Routledge
2 Park Square, Milton Park, Abingdon, Oxon, OX14 4RN

Simultaneously published in the USA and Canada
by Routledge
270 Madison Avenue, New York, NY 10016

Routledge is an imprint of the Taylor & Francis Group, an informa business

Typeset in Times New Roman by
Florence Production Ltd, Stoodleigh, Devon
Printed and bound in the UK by
MPG Books Group

British Library Cataloguing in Publication Data
A catalogue record for this book is available
from the British Library

Library of Congress Cataloging in Publication Data
Denisova, L. N.
  Rural women in the Soviet Union and post-Soviet Russia/
    Liubov Denisova; edited and translated by Irina Mukhina.
      p. cm. — (Routledge studies in the history of Russia and
Eastern Europe)
  "Simultaneously published in the USA and Canada"—T.p. verso.
      Includes bibliographical references and index.
    1. Rural women—Soviet Union—History.   2. Rural women—Russia
(Federation)—History.   3. Rural women—Soviet Union—Social
conditions.   4. Rural women—Russia (Federation)—Social conditions.
5. Women peasants—Soviet Union—History.   6. Women peasants—
Russia (Federation)—History.   7. Women—Soviet Union—Social
conditions.   8. Women—Russia (Federation)—Social conditions.
9. Soviet Union—Rural conditions.   10. Russia (Federation)—Rural
conditions.   I. Mukhina, Irina, 1979–   I. Title.
HQ1662.D46 2010
305.40947′091734—dc22   2010021317

ISBN10: 0–415–55112–9 (hbk)
ISBN10: 0–203–84684–2 (ebk)

ISBN13: 978–0–415–55112–0 (hbk)
ISBN13: 978–0–203–84684–1 (ebk)

To my daughter Maria Denisova
—L. Denisova

To my husband Denis and my son
—Andrey I. Mukhina

Contents

Figures

Tables

Preface

As someone who has been born and raised in the Soviet Union but now lives in the United States, I have always felt there is something special about going back to Russia. Predictably, I enjoy spending time with family and friends there as well as traveling around Russia to enrich my understanding of the Russian past and hence my own teaching of the subject matter. But above everything else, I always look forward to intellectual discoveries that enrich my own research; I crave the experience of finding new works of scholarship and meeting authors of these fascinating new accounts in Russia. On one of such trips about 5 years ago, I came across a newly published study entitled *Zhenshchiny russkikh selenii* (*Women of Russian Villages*; Moscow: Mir istorri, 2003), which dealt extensively with labor conditions and working lives of women in the Soviet countryside. I was immediately impressed with the scope of the book and the amount of research that went into producing it. "What a fine work!" I thought to myself.

Hence I was thrilled to get to know the author of the work, Dr. Liubov Denisova, only a few months later. Once our paths crossed, we kept meeting over and over again at various venues that made our academic world seem small. Liubov and I shared many discussions related to gender studies, especially in light of my own preoccupation with gender roles and national identities among ethnic minorities who were subjected to wholesale or partial deportations in the Soviet Union in the 1940s. During one of our meetings, Denisova told me the exciting news: in 2007, Rospen, a leading Russian publishing house, released her new comprehensive volume on Russian rural women, entitled *Sud'ba russkoi kresitanki v XX veke* (*The Fate of Russian Peasant Women in the 20th Century*). This extensive, 500-page book investigated in depth the gender roles in rural families, women's perceptions of their marriages and divorces, motherhood and educational opportunities, love affairs and rumors. While reading this volume, I could not help but be awed by how much these rural women's experiences were different from their urban counterparts' and by how inclusive this new study is.

In one of our numerous conversations, I told Denisova that I was disappointed that nothing of equal scope and richness has been written recently in English. Sure enough, the recent years have seen the publication of excellent studies that dealt extensively with women in Russia. Yet, their numerous merits notwithstanding, none of the recent works presented a *complete* history of specifically *rural* women in the *twentieth* century, as most of them either encompassed the entire span of Russian history or were edited volumes with essays on a diverse range of topics.

It was at this moment that a dream was born to write a book that would encompass the entire wealth of information presented in the two volumes, yet in an abridged and accessible way and in English. Little did I know that I had a daunting task ahead of me, not one of straightforward translation but one of fully re-creating a world of Russian women in the Soviet countryside.

The task of revising, editing and translating the two volumes went beyond basic editorial and language skills. Many assumption and generalizations that are taken for granted in a Russian-language scholarship needed to be explained and substantiated for an English-language audience that is accustomed to different standards, both in scholarship and in real-life experiences. No scholar in Russia needs to explain why Soviet people picked chamomile and greater celandine, or why rural women complained of having nice leather boots instead of felt boots in local village stores. But these notions and categories, along with a myriad of similar examples, are often unexplainable for English speakers and do not form an integral part of their common knowledge. As a consequence, it was not enough to translate the text; even after significant editorial cuts and structural changes, nearly every page had to be rewritten to meet these new standards.

Moreover, deciding what to include and what to leave out was a challenge every step of the way. It was unfeasible and unnecessary to write a 1,200-page, comprehensive history of rural Russian women that would include both Russian-language books in full. Yet we ended up shedding intellectual tears over letting numerous parts of the book and various topics go. In the alternative universe where most readers love to work their way through 1,000-page-long scholarly accounts, we would have included much more on labor conditions of rural women, on trade unions, collective farm codes, production rates, agricultural techniques, and power tools. But bearing our limitations in mind, we aimed to produce social history at its best, where our readers would be allowed to journey into an alternative cultured space and a different gender zone than the familiar world of today.

Structurally, the book is divided into thematic chapters instead of following a basic chronological narrative. Our goal was to anticipate all kinds of readers and their interests. Those who are interested in reconstructing the complex world of peasant women's experience in full, should most definitely read the work in its entirety. But there are always scholars, students, and general readers who are interested in a theme or two. For someone who is interested in the rates of alcoholism or prostitution or childbirth, for example, such structure allows to open a relevant section of the book and get a general overview of such themes in an efficient and inclusive manner. Thematic structure does not change the scope of the book but hopefully makes it easier to work with for those interested in specific themes or topics for their own reasons.

With this project, both Denisova and I have acquired many intellectual and institutional debts. We acknowledge those in other parts of the book, but my special words of gratitude go to my own institution, Assumption College, for appreciating and understanding the challenges and limitations of our workplace, and to my colleagues in the department for their collegiality and unfailing support. Their role is immense, even if they do not always realize it. To you, my fellow comrades! Cheers!

Acknowledgments

The help and friendly advice from my colleagues as well as their unfailing support are instrumental to this project. Special thanks go to my daughter Maria Denisova, who reviewed the legislative aspects of Soviet history for accuracy and consistency; to Professor Olga Vasilieva, whose advice allowed me to better understand the role of religion in Russian rural communities; to Professor Theodor Shanin of Manchester University for his kind permission to use unpublished materials of the British-Russian research project on Russian villages; and to Lynne Viola and Wendy Goldman for their friendly support. Roberta T. Manning actively participated in this project by reading various drafts of the manuscript and making suggestions for improvement, all of which were taken into consideration. Special thanks also go to Dasha Balaban. But as always, all errors, omissions, and misinterpretations are solely mine.

My appreciation and gratitude go to Professor Irina Mukhina of Assumption College, the organizer of this publication and the driving force behind it. It is only thanks to her initiative, energy, enthusiasm, help, and skill as an editor, translator, and at times co-author that this book came into existence.

Introduction

Russian women are the soul of the Russian village. The agricultural reforms that the Soviet government promoted at various times affected rural women more than any other inhabitants of the Soviet state. They were the first villagers to respond to the government's propaganda, by offering their emotional support for reform; and as the main labor force within the village, they were also those who carried out the reforms and, hence, were also responsible for the reforms' successes and failures. In the world of a Russian village, women remained the source of stability, perseverance, patience, and hope for the entire village community. Rural women always worked hard and saw this work as an essential part of their lives. In fact, work defined women's existence, as it took up the lion's share of their time and was both an expected and a required component of their daily experiences. Yet women's primary concerns and efforts were aimed at family, at their children and grandchildren.

An ordinary rural life of an ordinary rural woman was a long chain of everydayness, interrupted only by occasional festivities of the rural community such as weddings, births, and religious and secular holidays. A rural woman's dreams were often unfulfilled and her life was that of a hardship. Hence she aspired to create conditions for her children that would allow them to realize their potential and see their dreams come true. A rural woman attempted to reconcile the hardships of living off the land with those few and limited pleasures that she was seeking and that were available to her. More often than not this quest for reconciliation and balance failed, and the hopes vested in one's children became tied to faraway places such as urban centers and even the capital city, Moscow. Children also saw that their reality fell short of their dreams and ambitions on many occasions, and these children in faraway places could only miss the "small motherland" of their village and their parental home. Ambitions of rural women were not about exotic destinations and luxurious possessions; they included hopes of being able to live a decent and dignified—even if hardworking—life; of having a chance to raise children and grandchildren nearby; and of being close to nature and to their own rural community. These hopes for a decent but honest life of work and family are part of the rural life to the present day.

*

Women's history has been the subject of many scholarly studies, both in Russia and outside of it.[1] Most scholarly works investigated three main aspects of a woman's life, namely employment, family, and everyday lives. Thus, research that has been done in the last few decades addressed issues of women's participation in the labor force;[2] the challenges of combining domestic life and public employment;[3] everyday experiences of women;[4] demographic changes and trends in the Soviet countryside;[5] and the history of family in the Soviet Union and beyond.[6]

If the earlier works investigated women's history, new studies that emerged in the post-Soviet decades emphasized gender and gender roles as key categories of historical analysis. Such gender studies often have a clear agenda of using and applying their knowledge towards creating a society that is truly egalitarian and open to women. Not just telling women's history, but assessing the social construct of gender roles is the key to reconstructing a complete picture of women's experiences in the Soviet Union and elsewhere. All facets of women's experiences matter in this analysis of gender roles, including women's participation in the sphere of politics; women's employment rights; gender stereotyping and its role in enforcing the practices of everyday life; social protection and care extended to families; and even the perception of women and their roles in various scholarly disciplines such as sociology, economics, philosophy, psychology, and philology.[7] The international scholarly community has been instrumental in shaping the field of gender studies as applicable to Russian history,[8] and courses on gender are currently offered at most universities across Russia.[9] Equally, various studies have attempted to trace legislative changes that affected women and how these legislative acts have changed and evolved as a result of recent efforts to bring Russian legislature on a par with world standards and practices.[10]

This book is about the life of a Russian rural woman. The emphasis on a Russian woman is justified by the fact that Russians predominated both statistically and territorially in the Russian Federation (as a part of the Soviet Union) and in Russia over the last two decades.[11] Women in the countryside shared experiences that were at times common to their urban counterparts and at times drastically departed from the latter. At the time of the collectivization campaigns, Stalin acknowledged in his propaganda speeches that women in the collective farms were a major force and that peasant women needed to realize their labor potential. Indeed, over the next several decades, women were in the majority in the countryside, and worked both in the fields and husbandry as well as in teaching and medical professions. Yet collectivization did not create circumstances that would have allowed women to become "a major force" among the top echelons of power, as leaders and chairmen of the collective farms. Most women still occupied unskilled positions with low pay and long hours, all the while investing extra labor into their private gardening plots and into performing household chores.

Anti-religious campaigns that aimed at creating a new Soviet person devoid of attachments to "superstitions" such as religion drastically challenged the spiritual world of a peasant woman. The stronghold of the Russian Orthodox Church (ROC) was in the countryside and among women, and even prolonged struggles to root out this "opium for the people" did not eradicate religiosity and religious faith in the countryside. Furthermore, the Stalinist program of social transformation was heavily restrictive for women. This was confirmed constitutionally in the Abortion Ban of 1936, which remained in place until 1955, and in the Decree on Marriages and Family of 1944, which recognized the legitimacy only of those children born to parents in a registered marriage. Many children were marginalized overnight by this decree as those born out of wedlock and thus became no longer entitled to any compensation from the state or child support payments from their fathers. For rural women, the transformation of family affairs and religious life were of major consequence, disproportionally more important to them than to urban dwellers and especially men.

The Stalinist era ended in 1953, but the subsequent *Thaw* was of little meaning to most rural people. Though the state became more people-oriented, changing its economic, political and social priorities towards addressing the needs of its citizens, Nikita Khrushchev did not reverse the Decree on Marriages and Family of 1944 and did not extend any social benefits to rural residents. Pensions, social security benefits, a right to a passport, and subsequently a right to movement and freedom of employment—all these were denied to the majority of Soviet people who were still predominantly rural. The changes came only in the second half of the 1960s.

The Brezhnev era came with its share of pros and cons for the Soviet countryside. Indeed, the government passed a series of laws on family and marriage in the late 1960s and into the 1970s that improved social protection of motherhood and childhood. Moreover, collective farmers were granted the right to have a passport, and access to complimentary secondary and post-secondary education became truly universal. Yet at the same time, anti-religious campaigns intensified, compared to the previous two decades, once again attacking the core of rural life.

The Gorbachev era saw a complete restructuring and eventual collapse of the Soviet system. One of the unforeseen consequences of this transformation was a massive demographic crisis, in which the birth rate dropped well below the reproduction levels for many years to come. The hopes of a prosperous and easy life were not fulfilled, as the 1990s brought with them hyper-inflation, the demise of the economic sector, and a crisis of agricultural production, among many other problems. The loss of social benefits and protections extended by the late Soviet state was often perceived as a personal and collective tragedy, and new-found freedoms did little to compensate for this loss. For most rural dwellers, guaranteed employment, free education and medical care, free daycare and afterschool programs, and annually distributed free passes for family vacations in the Soviet resorts were more meaningful than their newly acquired right to establish transnational corporations or the right to multi-party elections.

*

This book relied extensively on archival information, especially documents found in RGASPI (Russian State Archive of Social and Political History), RGANI (Russian State Archive of Modern History), and GARF (the State Archive of the Russian Federation). Regional archives from Vologda and Archangel supplemented materials found in central archival collections. Social studies, and especially a project undertaken by a team of British and Russian scholars in the 1990s who interviewed residents of thirty Russian villages, allowed the book to build on a rich pool of sociological research previously unseen or unpublished.[12] Ethnographic studies and folklore were used to enrich the story and add a personal touch to political transformations experienced by the entire nation.[13] Regional and Moscow-based magazines, journals, and newspapers became an excellent source for the book as well. These sources were instrumental for writing the book, as well as for recreating an uneasy but fascinating life of a Russian peasant woman.

Part I

Employment patterns among rural women

1 Women's work

An overview

The twentieth century saw three significant eras of radical reform of the agriculture that drastically changed the lives of peasant women. The first significant change came with the Decree on Land of 1917 that stipulated an equal redistribution of land among peasants (and by default also the confiscation of land from those who had more than the allowed norm). The second change came with the forced collectivization during the campaigns of the 1930s, which were marked by the significant use of force and the destruction of prevailing lifestyles, and even mentality, in the countryside. The final drastic change came in the midst of the Soviet Union's breakup and post-Soviet transformation in the early 1990s. The final campaign sought to alter the Soviet system of agricultural production that was deemed "failed." Yet this was precisely the system that had become familiar to collective farmworkers who were born and raised into it. At the very least, these rural residents became accustomed to and counted on an extensive social security network that existed in the last decades of the Soviet power.

The most dramatic change of the three, undertaken by Stalin in the 1930s, was immediately labeled by the state as "a revolution from above, [also] supported from below." The goal of the economic reforms of the 1930s was to achieve the modernization and rapid industrialization of the country, but in order for this to happen the state had to rely on the village (for example, to finance industrialization by producing agricultural surplus for export). The interests of the rural population were never taken into account throughout the Soviet era, but the collectivization campaigns were unmatched in their brutality and in excesses that the campaign permitted to its individual players. Nearly 15 million people became victims of repressions in the later 1920s and into the early 1930s, and the repressions affected every third family in the countryside. The horrors and abuses of the collectivization campaign that claimed 8 million lives (most of which were rural),[1] were "not to be forgotten till death," as a famous Soviet writer Mikhail Sholokhov wrote to Stalin.[2] Typical of all Soviet repressions, the government classified all people according to distinct categories, and those people who were labeled *kulaks* of the first category (presumably the wealthiest peasants) were sentenced to death. They added their share, estimated at almost 400,000 people, to all death sentences carried out in the Soviet 1930s.[3] Kulaks who were assigned to the second category, were exiled to faraway and underdeveloped regions of the Soviet Union in order to be "educated by labor." From the 1930s to 1950s, of 6 million people who were sentenced to "re-education by labor," 2.5 million were a part of the kulak deportations.[4] But the state-sponsored repression against the rural residents continued even after the end of the collectivization campaign; in fact, repressions existed in one form or another throughout the entire Soviet era.

The pronounced Communist Party doctrines and resulting campaigns systematically affected the Soviet countryside more than Soviet urban centers. The repressive policies of the state were, of course, applicable to all Soviet citizens. But there were always special resolutions or decrees just for the rural dwellers who until the 1960s still comprised the majority of the population. Also most reforms that aimed to "build the socialism in one country" depended on the successes or failures of these reforms in the countryside. For most of its history, the Soviet agricultural sector produced food and supported the urban residents for minimal to no pay, and villagers also had significantly fewer rights and provisions than other Soviet citizens. Thus, again up until the 1960s, collective farmworkers had no passports (which were required for travel, relocation, and employment); no social security systems; and no guaranteed wages or incomes. They worked for the state, which often used the free and nearly free labor of its agricultural workers to achieve its grandiose goals. But even villagers called themselves "the second class citizens" and *derevenshchina* (a derogatory term for "uncultured villagers") who could get by in their "less than perfect lives as long as they could somehow make the ends meet."[5]

From the first days of the universal existence of the collective farms, every aspect of rural life was regimented by the three Codes: the Code of 1935, the "socially-oriented" Code of 1969, and the "democratic" Code of 1987. All everyday questions and concerns were also spelled out in the Rules for the Internal Regulation of the Collective Farm.

The state initiative to build the rural life around collective agricultural production was indeed supported by some villagers, or specifically by the rural poor, who saw an opportunity to improve their lot at the expense of better-off neighbors. Many kulaks who were deported and at times executed suffered not only from the Soviet decrees; they were also victims of envy of their neighbors and were singled out by the community itself, which despised these people's successes. The former "wealthy" peasants learned to live in remote steppe regions; they survived and at times even fared relatively well in the context of their unfortunate lot. But the remaining rural people who were not singled out in the collectivization campaigns faced unusual challenges. They "inherited" all the property—land, houses, mills, barns, along with equipment and livestock that these kulaks possessed—yet they were poorly prepared to handle it and to do so effectively. For most parts, the people who remained had little practical experience of running large households; they were typically hired by those wealthy kulaks who had the knowledge and the expertise to oversee the farm work but who were now exiled to Siberia. At best, the remaining peasants had small plots that they used to grow grain and foods for personal consumption; after all, anyone who had had more was already gone. But the state also saw no need for a personal initiative from these peasants, as they were supposed to be united by a common socialist goal and a common collective farm existence.

These collectivization campaigns were the starting point at which the village was systematically altered and even pillaged, most importantly of its manpower. Facing the growing shortage of manpower, Stalin started to praise the female collective farmworkers, labeling peasant women "the great force in the collective farms" on more than one occasion. Women were in the majority in the countryside, and they also performed the bulk of the manual labor on which the agricultural production depended so much. Stalin wanted to rebuild Soviet agriculture on the back of this predominantly female labor force, and to do so by first cleansing it of its "religious opium." But in practice, with the exception of young and radically predisposed female party workers who internalized all Soviet ideology at face value, peasant women could not and did not want to embrace the numerous reforms of the Soviet state. *Bab'i bunty* demonstrated many times over again peasants' preferences for individual household farming. The village turned to collective farming only because of the repressive measures of the Soviet state.

For the Soviet Union as a whole, female employment rates were high, as over 90 percent of all women worked outside home. In the countryside, women constituted at least half the workforce, and at times even more. The Soviet Union had one of the highest rates of female employment in the world.[6] Yet at the same time, as subsequent chapters will demonstrate, women were solely responsible for all or the bulk of the household chores and child rearing. Even though the social services, such as nearly-free daycare centers and common eating facilities, were meant to alleviate some of these hardships, these services were not always universally available, and even when available they never completely substituted for a woman's role in the family. As a consequence, women worked long hours due to a combination of their *double burden* of public employment and domestic duties, and a rural woman often worked more than her urban counterpart.[7]

Women remained the main labor force and the backbone of the countryside during the entire Soviet period, in terms of both agricultural production and the social aspects of life. Yet the role women played in everyday life and the leverage they had in the decision-making process, as well as informal power that they welded in the community, often depended on such factors as male outer migration and subsequent sex-ratio imbalance (more women, more power); an individual choice of redistributing time between the collective farm work and household chores; the type and level of education and training that these women had; and the state policies, especially those aimed towards the protection of motherhood and childhood in the countryside.[8]

Peasant women worked equally alongside men in the collective farms and systematically outpaced men in terms of the labor days that they worked (a common measure of employment in the Soviet days) and in terms of human hours that they invested. It is important to note here that rural women rarely if ever assumed leadership positions and were under-employed in skilled jobs that implied higher training but also higher pay. Instead, they were routinely employed in low-paying manual jobs, especially in animal wards and milking where women represented nearly the entire workforce. This division of labor along the gender lines persisted throughout the Soviet era.

The public opinion that the rural, collective farm life was "of second class" was created and promoted by those former peasants who managed to move to cities and remained there for life. New urban residents, who felt that they had made their dreams come true by landing the job of a tram conductor or a bus driver in a big city, did not want to return to their roots. They often spoke of their fellow villagers and rural life in derogatory and diminutive ways, thus reinforcing the image of a backward countryside. This image was so remarkably pervasive because the Soviet Union was on the move in its first decades and experienced massive migration from rural regions to urban centers. In that sense, the city was formed and shaped by the countryside and by all those former peasants that resented their prior lifestyle and their prior experiences. And even though some "kept dreaming of motherland at night, [and of] their village with which they are bounded in heart for life," many more were ashamed of their rural past.[9]

Other factors also reinforced such notions of backwardness. "Russian village" and "collective farm" were not identical terms that could be used interchangeably, but all people who lived in the countryside were still termed as "peasants." So some peasants loved their village and their traditional lifestyle but resented the collective farm system. Yet they had no alternative to the system, and because they resented it, they by extension and out of lack of choice came to reject all facets of the rural lifestyle. It is a tempting hypothesis to consider what might have happened to the Soviet village had the state allowed, even in limited ways, for traditional village life to survive. The collective farm often contradicted the mentality of the village life, and hence at least some rural dwellers felt as if they had become internal self-imposed exiles who could neither go back to collective farms nor find a replacement for their home in large urban centers. This internal craving for a rural lifestyle also helps explain why the outer migration, though substantial, nonetheless did not include all those who resented the collective farm system. Peasants could not part with their community even in opposition and protests to the state and instead opted to wait out until better times came.

This rural lifestyle had many unique features to it but primarily implied a slow pace of life during most seasons with minimal if any stress; it had wide-open spaces of the steppes and fields; and it was centered on a small, tightly-knit community that shared every joyous or tragic event of one's life. The community was a place where everyone shared a common set of values, a common set of concerns and difficulties, and where one could be understood without words. Rural life also left some room for people to feel like they were in control of their own lives; they were the heads of their households and masters of their own gardening plots, if nothing else. The fact that nearly all members of these village communities knew each other, were born in close proximity to each other, and grew up playing together, gave peasants an extra special sense of an extended family, with all its pros and cons. None of the above could be replicated in cities. That is why when the younger generation, born under the Soviet power, went to cities to study and stayed there for good, very few of their parents wanted to join them, even when the living conditions would have been better. Peasants who lived their entire lives in the countryside found that when it came to cities, "it is nice to visit, yes, but no, not to live there."[10] The sense of unity with the great outdoors was also significant as the rural life offered an opportunity to "live in harmony with nature."

The collective farm was a product of the Soviet system and went underwater along with it. The reform of village life that was attempted without people's understanding of what the reform entailed, did not solve the country's food problem. Instead, by undermining the producing sector, it exacerbated the problem and caused massive man-made famines. The collective farms created in the 1930s could not fulfill hyper-optimistic hopes for what they could produce. Yet they managed to level out every peasant to a mediocre average of socialist equality. Hence when peasants speak today about their nostalgia for the good Soviet days, they do not crave the collective farm. What they yearn for is the sense of security that the Soviet system made possible, thanks to its extensive social security web and various social guarantees.

For Russia, the agricultural question was always of major importance, not the least because the majority of the population up until the 1960s lived in the countryside. Even at the beginning of the twenty-first century, a third of all Russian citizens continue to live in rural parts of Russia, a percentage that is high relative to most other European nations. Back then and now, the majority of the adult rural population are women. Moreover, as the later sections will expose, the bulk of the agricultural production in present-day Russia is still carried out by women. Over half of all agricultural produce comes from family farms and private plots, and the labor in both is predominantly female. There are strong indicators that this situation will persist into the future as well, at the very least because the male life expectancy is 57 years while female life expectancy is higher by 15 years. Because of their skills and amount of labor invested in farming, peasant women represent a substantial voice in the village life.

Family farms as they exist now are not uniform as they perform a wide variety of tasks and produce diverse agricultural products. Family farming systems became appealing to women because they had to perform household chores, raise children, and do most jobs that came their way, while family farms allowed women to combine their responsibilities with farming done right in their backyard. Provided that the government continues to support private farming at or above the current levels, this system might prove to be the most viable option for the post-Soviet agricultural sector. Family farming is likely to be even more successful if the state acknowledges the special role played by peasant women. Women do a large share of work, of course, but they also preserve a lifestyle that so many villagers found irresistible back in the early twentieth century. Statistics demonstrate that thousands of villages have disappeared in the Kostroma, Yaroslavl, Tver', Pskov, and Vologda regions in recent years. Along with these villages goes a lifestyle as well. Nonetheless, in many villages women take control and, without waiting for state support or help from their male partners, start their own family farms and run successful businesses. Hopefully, these women will manage to preserve the customs and traditions that for decades or longer shaped the rural lifestyle of so many Russian citizens.

2 Unskilled labor in the countryside

Nothing got without pains but an ill name and long nails.[1]

Prior to the October Revolution of 1917, peasant women were involved in all farming and domestic chores. Girls began to help around the house and in the fields at an early age, and women typically did all kinds of jobs that were essential to provide for the family. Labor was divided along the gender lines, with the heaviest work of turning the soil assigned to men. But none of the work was divided into public employment or the domestic sphere; in the system of near-complete self-sufficiency, all work was done to shelter, feed and clothe the family.

The new system of collective farms that the Soviet government introduced and enforced in the 1930s changed the lifestyles of peasant women. In addition to their work around the house and in animal husbandry, women were also required to work for the collective farm. The codes that regulated daily life on the collective farm also required both women and men to volunteer for the good of the community by working on communal projects. The failure to satisfy these requirements implied severe sanctions, mostly in the form of prohibitions to use any pasture for personal livestock, confiscation of private gardening plots, and even resettlement away from the village to the remote parts of the country. Eventually, women in the Soviet countryside were represented in all jobs and professions, including agronomists, veterinarians, tractor drivers, and mechanics. Yet they remained the predominant labor force in jobs that were unskilled, manual, and low-paid, and the only labor force in jobs such as milking cows. In jobs that required heavy manual labor, women's output was only about 80–85 percent of that of a male worker. Hence the use of female labor was not the most efficient for such jobs.[2] However, nearly all occupations in farming and agricultural production that involved heavy manual labor were exclusively staffed by women.

Unlike grains, the production of raw cotton, flax (linum) and sugar beets was not mechanized in the Soviet Union and was a time-consuming and labor-intensive process. Even when the harvesting of flax was mechanized, on most occasions the machines were not functioning at all or were performing poorly.

Thus, flax-harvesting machines, even when functional, did such a poor job that women went for a second round of harvesting the same field to pick up what was left after the machines. The monotonous manual work, picking raw cotton and flax, was performed exclusively by women. The harvests of the 1930s were high, and the government had a good reason to reward women who were involved in growing such crops.[3]

Unlike industry where most overproducers (called *Stakhanovites*) of the 1930s and the 1940s were men, over 80 percent of overproducers in the agricultural sector were women, and their share was the highest among field workers.[4] The government policy was intentionally cynical when it came to these women; on one hand, the government praised and rewarded women who performed well or even outdid their daily quotas. Yet on the other hand, it failed to note that male workers were equally capable of producing as much or even more, but these male workers were specifically and intentionally assigned other jobs that were mostly mechanized. Stalin emphasized the important role women played in the agricultural sector. "Take, for example, Mariia Demchenko, a famous sugar beet worker," he remarked about the Stakhanovite movement in the collective farms. "She got record harvest of sugar beets of 500 quintals [centners] and more per hectare.[5] But can we use it as a standard measure for all harvest in, let's say, Ukraine? No, we cannot."[6] However, this did not mean that men were incapable of investing an equal amount of hard manual labor. It implied that men were not expected to do what was perceived as predominantly female field work. According to the party's vision, women had to understand their new place in the new society and follow the party's course, at times even becoming the avant-garde of this new path to the bright future.

Mariia Demchenko, who was praised by Stalin, got her first job back in 1930 in one of the Ukrainian collective farms. For her active role in a local Youth Communist League (*Komsomol*), she was rewarded by being appointed as a leader of an all-women field-working brigade. In 1934, she organized a brigade of eight young women to grow sugar beets. They were given 4 hectares (approximately 10 acres) of land to work. All work was done manually, with the sole exception of soil turning. The young women wanted to meet and exceed what was expected of them. In the wintertime, they started going around collecting coal and ashes, to be used later as fertilizer. When the spring came, they manually spread the ashes to fertilize the field. Then they manually pulled the weeds out one by one. When the summer came, they had to water the field by bringing buckets of water from a local well, once again with everything done by hand. Mariia recalled:

> Down the hill on one side of the field there was a spring. And seven days a week, we brought buckets of water to water the beets, for five hours non-stop, from five to ten at night. [After that,] we could not feel our hands, we had bug bites all over our faces; hands, feet, everything hurt.[7]

Figure 2.1 Women, members of the collective farm "Matveevskoe" planting cabbage sprouts, May 1940, Russia, Moscow region, Kuntsevsk district.

Source: Photo taken by Plotnikov. © Courtesy of the Russian State Documentary Film and Photo Archive, Krasnogorsk, Russia

In 1934, these young women, headed by Demchenko, harvested 469 centners of beets per hectare and 524 centners in 1935, while the national average was only a third of this number (150 centners per hectare).

By 1938, over a half of all 1,500 women rewarded with highest national honors such as the Lenin's Award and the Red Banner Award, were recognized for achievements in the agricultural sector.[8] Young women who looked physically attractive were preferred as symbols of socialist achievement over their older counterparts; these young women made literally beautiful examples of what every woman should want to be.

Some women were trusted to work with machines and did fantastically well, although in general most machines were rustic and primitive, and a good deal of manual labor was still left to be done. Thus, one worker, Manefa Siteneva, worked with a flax-processing machine. On average, a worker operating this machine produced about 25 kilograms of linen thread, whereas Manefa managed to have 1,064 kilograms. She explained that her exceptional output was a result of

> proper care for the machine, and testing to see what was the best spot to be for a worker next to the machine, what hand movements to use when grabbing the flax, how to pass the leftovers on to a helper.

Of course, she also added that the main part of her success was "a socialist attitude to work ethics and socialist enthusiasm." She was invited to Moscow to be rewarded for her achievements.[9] Typically, women were the most likely to be rewarded for work that had been considered *female* prior to the October Revolution. Thus, certain stereotypes carried into the new reality of self-pronounced socialist gender equality.[10]

Women who were recognized for their achievements greatly appreciated this recognition. But their neighbors were often less welcoming and at times outright hostile to such overproducers. Rudeness and threats were common, but neighbors also resorted to breaking windows in the houses of these women, or taking their own cattle to these women's gardening plots, thus destroying all their hard labor. Documents also record cases of arson, poisoning of cattle, and vandalism in the fields where these women worked. Villagers believed that the communal life that was typical of a Russia village in pre-revolutionary days implied a rough equality among the community's members, and that women who earned medals were show-offs who had no place in their communities.[11]

The Soviet press often reported such cases and, to use the jargon of the time, used them as examples of class struggles or cases of revenge by enemies of the people and alien class elements. Local collective farm officials, though, tried to ignore such cases as much as possible. These officials shared into the

Figure 2.2 M. Slepysheva (in the front) and her brigade in the collective farm "Smychka" during hay harvesting in 1944, Russia, Kemerovo region.

Source: Photo taken by Fateev. © Courtesy of the Russian State Documentary Film and Photo Archive, Krasnogorsk, Russia

common opinion that women should not show off their achievements, and collective farm chairmen tended to minimize or decrease the days worked by such overachievers, to falsify the amount of milk their cows produced, and to deny these women hay to feed livestock. On occasion, chairmen even sold cows that were under the supervision of the most successful women, believing that the awards, prizes, and any possible monetary compensations (though few and far between) were not for women alone but to be shared by all, and especially by chairmen of such "perfect collective farms." There were also concerns that these women could be promoted and eventually replace the chairmen.[12]

By 1937–38, women who topped the list as Stakhanovites of agricultural production, including M. Demchenko, P. Angelina, P. Kovardak, and M. Globa, were admitted to the top agricultural colleges in the nation. Moreover, they received as prizes material possessions such as bicycles and phonographs that were beyond reach of nearly all other people, except for a few members of the party elite. These women were also taken to Moscow to see Kremlin and party elites, and such trips were out of reach to most people. The attention these women were showered with was akin to that given to cinema superstars and politicians. Their lives were an object of envy by many, and the lifestyles of the few Stakhanovites who achieved such glory and fame were indeed unmistakably outstanding.[13]

Even when Stakhanovites did not receive such superstar treatment, their achievements and awards were enough to stimulate many women to do their best in overproducing. The attention paid to these women was a new phenomenon in a village life. These rewards also allowed the government to pay little attention to mechanization of the harsh labor that the peasant women performed; awards and prizes were sufficient in themselves to improve and stimulate production and even overproduction. Manual labor in the fields and in the animal wards remained a symbol of a female rural life up until World War II, even in sectors where tools and machinery had already been invented and could have been made available under different circumstances. In 1940, two-thirds of all 31.3 million women of working age in the Soviet countryside were involved in doing manual tasks for collective farms.[14] Folk songs and rhymes reflected this well, and one of them went as follows:

> I am a horse; I am a bull: I am a woman; I am a man. I sow the seed and I harvest, and I even carry the firewood on myself.[15]

The onset of World War II dramatically changed the gendered division of labor in the countryside. Women became the main workforce that had to substitute for men at all levels of professional and skilled occupations. Women became collective farm chairmen, agronomists, veterinarians, mechanics, tractor drivers, although they continued to work hard in manual jobs where machines were unavailable. Even in jobs that were mechanized, once tractors and other power equipment broke down there were no resources to fix them, as all resources were diverted to support the war effort or simply because there were no skilled mechanics to carry out the repair. Women recall how, during the war, they even had to turn the soil manually with nothing but shovels, and used their collective force to pull a plow, when "eight or ten women get to pull the plow together . . . as one teenager and one more woman was walking behind the plow."[16] Harvesting was also progressively becoming increasingly manual during the war as machinery became less and less available. During the harvesting season, women never left fields and preferred to rest right where they worked. Horses could not handle the demands of the labor and were spared the toughest jobs; moreover, each horse only worked a half-day shift. Women, however, could not afford such luxury and had to work on most occasions 18 hours a day, without getting necessary rest or food provisions. Only 2 percent of tractors were spared from the now-occupied territories, and even these were not in the best working condition. As a consequence, much harvesting had to be done with sickles and scythes. Women moved harvested grains during the night to optimize their performance; they needed to resume harvesting once the sun was up. This around-the-clock labor continued throughout the harvesting season. Women managed to harvest everything in non-occupied territories in 1941. The harvest of 1941 lagged behind the interwar harvests, but not significantly so. If in the interwar years the average was to take 8.4 centners of grains per 2.5 acres, in 1941 the average was 7 centners per 2.5 acres. The overall plan and goal of agricultural production in 1941 was to produce 80 percent of that in the interwar years.[17]

These numbers, although lower than previous years, were perceived as a true victory of female collective farmworkers in a war-torn country. For the duration of World War II, women remained the main and often only workforce in the countryside. Women worked the land with or without machinery, and supplemented their work by knitting woolen clothing and socks for soldiers, and writing letters to their loved ones at the front lines.

When surviving soldiers returned home at the end of the war, they found more than just their loved ones. They also discovered that many farms were in desperate need of maintenance and that women had often supplied the front lines with food at the expense of their own rations. Women plowing the land and using their own force, instead of animal labor, could not be sustained much longer. The village was already exhausted to the limits. The famine of 1946 also exacerbated the problem. In the immediate post-war years, most planting and harvesting was done, once again, with the use of machinery. But there remained many tasks that could only be performed manually and these were done by women. Much field work and care for orchards and vegetable gardens was still not mechanized. These jobs were assigned to female collective farmworkers, who carried out various tasks depending on the season and region-specific produce.[18] Even in 1970, two-thirds of all work in the countryside was unskilled and manual, and this proportion did not change for many years.[19] When it came to manual tasks in field work, on orchards and gardens, 98 percent of all jobs were done by women.[20]

In the fields, women mainly sorted the seeds in the wintertime to separate bad seeds from seeds suitable for planting; they de-weeded soil in the spring; and dug out potatoes, carrots, and beets when these were ready. Women were also actively involved in growing raw cotton, rice, tobacco, and all fruits and berries. Especially tiresome was work on vineyards. Each woman was typically charged with the care of 2,500 grapevines, which had to be trimmed and sprayed; the soil around them had to be turned and then fertilized several times; and grapes had to be cut off manually and carried on women's backs across the entire vineyard to a common storage shed.

Unskilled agricultural work was also seasonal and low-paid. The wages were even lower if recalculated on an annual rather than monthly basis; these women were laid off during the winter months because of the seasonal nature of their work.[21] Most women who did these jobs were older married women with no skills and no education, as well as young mothers who found these jobs suitable to their schedules and demands of motherhood. None of them were satisfied with their pay or the harsh demands of the job. But for some women in the countryside, there was no alternative to make their ends meet. Even young mothers who were entitled to child support payments from the

Figure 2.3 Komsomol youth from a collective farm "Zaria" on their way to field work in 1943, Russia, Kirov region.

Source: Photo taken by A. Skurikhin. © Courtesy of the Russian State Documentary Film and Photo Archive, Krasnogorsk, Russia

state received so little that they needed to supplement these state subsidies by working on collective farms.[22]

Unskilled labor still required a thorough understanding of the job and excellent attention to detail. For example, when seeds for beets were first planted, they sprouted in mass, and some had to be pulled out to allow for other sprouts to grow well. One woman who worked with beets later recalled

> that it was crucial to make no mistake and pay attention to what you did, though it demanded some patience. You had to make sure that only the strongest and best new greens stayed and weakest were pulled out. On average you get about thirty new sprouts per meter but you have to leave only six or seven there. We had to pull out about 400,000 new sprouts per hectare. Right after that, everything had to be de-weeded. My lower back hurt constantly, and my feet got very heavy.[23]

Growing beets remained a largely manual process, which also demonstrated well the gender division of labor in the Soviet countryside. All machines and power equipment were worked and serviced by men. All leaders of work teams and all collective farm chairmen were also men. Yet all manual jobs such as pulling the weeds out were done by women. As a leading Soviet journal *Krestianka* (a female peasant) reported, "men have machines but women have experience, natural patience and in-born skill to work with this tricky crop."[24] Folk songs and popular rhymes demonstrated well how women felt about such labor division:

> Our chief is good, as
> He does not do a thing.
> He only puts on a white dress shirt
> and poses around the village.

Ironically, the gendered division of labor was based on one's access to machinery more than on any perception of a particular job as masculine or feminine. That is why once a job or a task became mechanized and no longer done manually, it shifted from the realm of a women's responsibility to the realm of men's work.

There was also an emerging generational divide in the post-war era, especially starting in the 1950s. Young women who were guaranteed high school education by the state and received their high school diplomas, were reluctant to work alongside their older female relatives in manual unskilled jobs. These younger women also realized that much of the manual work could be done by machines if the state made sufficient technology and power tools available to all collective farms. They requested a different treatment for their labor and wanted to see the production become more up-to-date and thus fairer to women. One woman shared her sentiments during the so-called round table discussion published in *Krestianka*:

Because of poor mechanization, women who work manually play a crucial role in growing sugar beets. But what is the predominant age of these women? They are at least in their forties, though most are in their fifties ... Younger women with high school education do not want to do the job. We need better, 100% mechanization of the beet producing sector. And while we are dealing with getting technology to the countryside, we at least need to give more respect to these manual workers.[25]

By the 1970s, the government attempted to minimize its dependence on manual unskilled work by relying heavily on pesticides and other chemicals. Although the amount of manual labor presumably dropped, the use of chemicals raised significant safety concerns. For example, in Kichmengsko-Gorodetskii district of Vologda region, in 1979, a total of 22 farms stored and used chemicals. However, four of these farms had no special storage for them, and seven had sheds that did not meet basic safety requirements. None of the farms had a license to store or use these chemicals. There were no meetings or training sessions for farmers to explain how to handle potentially hazardous and life-threatening chemicals. Few if any workers had their medical exams, though according to law all workers had to have one. Officially, the full workday for those who handled poisonous materials was four hours a day for five days a week. But even this condition was nearly never met. Some women did not know about it, while others ignored it and opted to work eight instead of four hours to get double pay. Nearly universally, both men and women underestimated or did not fully understand the threats of working with such compounds, and they did not follow even basic safety guidelines for handling chemicals.[26]

The same story of insufficient mechanization that existed for sugar beet production was evident for other crops as well. As late as the 1970s and even in the 1980s, half of the entire annual potato harvest was dug out manually with shovels, and most of it was done by women. With other vegetables such as cabbage and carrots, women still manually pulled out weeds from the soil and harvested three-quarters of everything that was grown. The only way for the farmers and the state to handle the pressure of harvesting was to involve students and military personnel for seasonal labor. For example, college and university students were required to spend a month in the beginning of each school year working on assigned fields, or else they were denied their right to attend classes. Although this policy did relieve some pressure, the bulk of harvesting work still rested on women's shoulders. In 10 years, from 1975 to 1985, the share of unskilled manual labor done by female workers dropped only from 88 percent to 83 percent, thus still remaining high in absolute numbers.[27] Many women remembered that, along with their experience, they also "accumulated" back strains, prolapsed wombs, and other health problems.[28] Yet these women had few if any choices and continued to carry on with their heavy burden of unskilled agricultural work. When questioned by social scientists, only a few women stated that they disliked their jobs outright. The majority said that they were "fine with these jobs because there were no other jobs here."[29] Indeed, even for professionally trained specialists, the choice of training was limited to a few occupations essential to farming, such as agronomists and veterinarians.

By the 1980s, the growth of the industrial sectors allowed for greater employment opportunities in the countryside. Some villages now had textile factories, while others even had relatively high-tech industries such as radio equipment factories. Female collective farmworkers were only allowed to be released from their farms and assume these jobs under the condition that they would be available to perform unskilled jobs during the harvesting season. Only once the harvesting season was over were they allowed to go back to their jobs at the local factories. For example, in one village in Pskov region a new textile factory was willing to hire any local young woman with a high school diploma. Some young women even returned back home from their jobs in local urban centers so that they could work in this factory. The factory was of a high quality, and built to the best sanitary requirements; it had modern machines, good lighting, and proper ventilation. The main condition imposed by the collective farm chairmen, however, was that the factory sent all its female workers to work for a month or two at a local collective farm during the high season. In many ways, this and similar stories presented a win-win situation. Women gained greater access to better employment; factories had a steady supply of workers; villagers were pleased to keep their young people around instead of losing them to migration to cities; and collective farms still had their workforce during the harvests.[30]

In the 1980s the Leningrad region had numerous collective and state farms that cooperated with factories or even started their own production. Six farms opened their own factories where they employed over 500 people in manufacturing glass and paper containers. An additional 96 farms had agreements with 130 local industrial enterprises to hire women for off-season work. Once again, the main goal was to use female labor when it was not needed in the fields.[31]

Even when women had a few minutes of spare time, they continued farming on their private gardening plots. Although these gardening plots will be discussed in later chapters in great detail, it is important to keep in mind that women and only women were expected to work these plots, and once again women performed all the hard manual labor. The only difference was that this was a labor of love that allowed women to supplement their incomes by growing some of family's basic food ration. But the work itself was no less exhausting or tedious than collective farm work.[32]

In the post-Soviet 1990s, the Russian countryside experienced a massive economic depression and the loss of the social security system that existed in the Soviet era. These difficulties were paralleled by similar challenges in the cities. But women in Russian villages were the main source of optimism and at large displayed a positive outlook despite the grim realities of their lives. "I won't let things collapse; everything would be fine," assured one woman when she spoke with researchers from Moscow in the 1990s. But she added that "it takes so much energy, so much force to simply stay afloat."[33]

3 Female mechanics and machine operators[1]

All women, to the tractors!

Even though rural women constituted the backbone of agricultural work in jobs that were unskilled and manual, Soviet propaganda for gender equality and mechanization of the countryside brought changes to women's lives, at least for some. Stalin's slogan that mechanization would solve all problems implied a greater than ever need for skilled laborers capable of operating machinery. In the 1930s, the government set a quota for women enrolled in professional training schools. No matter the personal feelings of individual collective farm chairmen, 25 percent of seats in these professional training schools had to go to women. As a consequence, some women received training as tractor drivers, combine operators, and mechanics of all types.[2]

Women who worked these jobs faced the same problems as men and received no preferential treatment or privileges. They worked long hours, had no sanitary facilities in the fields, slept under the open skies during high seasons, and had no heat or water to use. All machines were also highly polluting, and produced high emissions, and no workers were safeguarded against poisoning from overexposure to fumes. In the first Soviet tractors, the force that had to be applied to turning a steering wheel was equal to lifting 30 to 40 kilograms (approx. 66 to 88 lbs). Accidents were common, and machines routinely broke down and required repair.[3]

All tractors were manufactured to be operated by men. Seats were uncomfortable and also positioned too far from everything else; as a consequence, some women could not reach the pedals or steering wheel because they were not tall enough. The People's Commissar for Field Work, M.A. Chernov, issued a directive to make seats adjustable and suitable for women, but this directive brought no changes and had no consequences.[4] Women continued to struggle in order to overcome the physical discomfort of working these machines, as well as the prejudice against them working in these jobs.

At first, women faced significant discrimination and prejudice from many men who felt that women should not operate or repair machinery. At the same time, official data demonstrated that women were better at working machines; they attended to them more cautiously, had fewer repairs, and worked more land on average than men. The first working brigade of female tractor drivers was organized in Donets Basin in the present-day Ukraine in 1933 and consisted of eight young women headed by Pasha Angelina. On the first day of work, the brigade was met with a demonstration of local women who threatened the young tractor workers and yelled at them to "Turn the hell around! We will not let women's machines on our fields!" But this brigade worked so well that eventually locals changed their attitude and "begged to forget the past! [since the demonstrators] were irrational and ignorant and this should never happen again." Women of the brigade worked hard to produce 1.5 times the state norm in the first year, and they even organized a vocational school to train young women to work tractors and combines. In 1936, their school enrolled over 100 young women.[5] Soon enough this brigade was known all over the country, and its example seemed worthy of imitation. In 1937, there were already 1,250 female tractor teams nation-wide, and half of them harvested more than double the state norm. Some of the best teams even covered 1,200 hectares of land whereas the national average was 372 hectares.[6]

But once again, local men created all sorts of obstacles for female mechanics. Collective farm chairmen routinely denied women assistance with preparing the fields for sowing, something that they did for male teams of workers. Some chairmen refused to provide female teams with food and water, and did not allow chefs to cook for these brigades. Equally problematic was the distribution of diesel and petroleum as male teams were given preferential treatment and were first in line to get these scarce but much needed resources.[7] Yet despite such hostilities and negative attitudes, the popular perception of women in "male" jobs started to shift in favor of supporting these women.

Rumors, however, were harder to overcome. Women who worked agricultural machines spent most of their day surrounded by men, and they even slept under the open skies together with their male co-workers during the harvesting season. Such proximity was the subject of many rumors and gossip about the paternity of children and fidelity of wives. Pasha Angelina, for example, complained that her mother-in-law always pointed out that Pasha's daughter looked nothing like Pasha's husband, even though there were no grounds for such accusations. Pasha's husband repeatedly asked her to quit her job to spare them such rumors, until one day he asked Pasha to choose between their marriage and her work. Pasha chose work.[8] All of this was going on at the time when Pasha had already become a national hero and a symbol of agricultural production and was awarded the highest honor of the Soviet Union, the Order of Lenin. In her years as a tractor driver, Angelina, the most recognized female mechanic of the country, received the highly prestigious status of the Hero of Socialist Labor twice; became a member of the Central Committee of the Communist Party of Ukraine; became the deputy of the Supreme Soviet of the USSR; was decorated with the Order of Stalin, three

Figure 3.1 Renowned tractor driver P.A. Angelina during harvest on the fields of the collective farm "Politotdel," 1941, Russia, Stalinskii region, Starobeshevskii district.

Source: © Courtesy of the Russian State Documentary Film and Photo Archive, Krasnogorsk, Russia

Orders of Lenin, and the Order of the Workers' Red Banner. Each order in itself was a highly prestigious award, but such combination made Angelina unprecedented. She was also given her own monument while she was still alive.[9] Yet all of this recognition did not spare her the sarcastic remarks and gossip about her "affairs" that eventually led to her divorce.

Working the machines was also hardly compatible with the demands of motherhood, both while pregnant and with small children. Female mechanics worked far from home, and their children had to be with them in conditions that were not suitable for small children. That is why two-thirds of all female combine and tractor drivers were under the age of 25, and the majority of them left their jobs once they got married and had children. The physical demands of the work were too high for young mothers to carry on with their jobs.

With the potential war looming on the horizon, in 1939 the government stepped up its efforts to recruit more women into skilled labor in the countryside. The government showered even more attention onto role models such as Angelina, and stepped up its propaganda campaign to have "a hundred thousand girlfriends on tractors." In 1939, 200,000 women had completed their training and worked as mechanics in various jobs.[10] The numbers of women working in these jobs in the Soviet countryside became statistically significant for the first time in history.

Figure 3.2 R.I. Liashenko, a collective farmer, behind the steering wheel of a combine during the harvesting of wheat in the fields of the agricultural *artel'*
"Po puti Iliacha," 1945, Russia, Krasnodar region, Labinskii district.

Source: Photo taken by Kal'nitskii. © Courtesy of the Russian State Documentary Film and
Photo Archive, Krasnogorsk, Russia

The coming of World War II dramatically changed the gender dynamics
in the Soviet countryside. With many men off to war, the collective farms
had to rely extensively on female labor to work the fields. Crash courses were
organized to train all volunteers how to operate machinery, and thousands of
enthusiastic people enrolled on those courses. During 1941–45, over 2 million
people were trained to operate and repair machinery, and over 1.5 million of
them were women.[11] Moreover, female teams on average produced more
than their male counterparts.[12] During the war years agricultural machinery
often broke down but was not replaced or fixed, and was overall insufficient
to satisfy the demands of the war-torn country. To compensate for these
deficiencies, women opted to work around the clock in two shifts of 12 hours

each to maximize the use of machinery that was still available and operational.[13]

Women who worked the machines during the war were often housewives prior to the war years and had limited or no experience with large-scale agricultural work. For example, Daria Garmash saw her husband off to the warfront in 1941, while she stayed behind with her 2-year-old daughter. She decided to organize a women's tractor brigade. She became a leading advocate for recruiting women to work tractors and combines, and decided to promote this initiative by the means of socialist competitions that awarded prizes to the best-performing teams. Unlike most other women, she stayed with her new-found passion, and after the war became the supervisor of Rybnovskaia garages of the Riazan Province. After 1961, Daria became a Deputy of the Supreme Soviet of the USSR and a winner of the State Award for improving the methods of use of wheel-supported tractors. Eventually, she became a heroine and a symbol of success for Soviet agriculture.

After World War II, women mechanics continued to play an important role in the agricultural production of the country. In 1956, 254,000 women worked as tractor drivers, combine drivers, mechanics, and leaders and chairmen of various female teams of workers.[14] Among them, there were always women who loved the "steel horse" more than anything else and could no longer perceive their lives without machines. Tractor operator A.I. Slizneva, evaluating her work, wrote:

> I work for the third collective farm already, and I am not boasting of my work, but nonetheless want to tell you that of the two tractor teams that we have, mine is in the lead, and we will not let this leadership slip away from us in the future. I want you to trust me that if I took it as my responsibility to do the work, then I will fulfill that responsibility fully; be sure of that. I will not complain of all the hardships that are ahead of us; rather, I will march toward them.[15]

The number of women who worked as mechanics started to decline once the Red Army experienced a mass demobilization at the war's end. Even Pasha Angelina, numerously rewarded for her work, had to give up the idea of leading a team of women. Personally, she kept her job and continued working on her "steel horse" up until the retirement age, but after the war her only option was to supervise a team of male workers.[16]

The number of women on the machines continued to decline rapidly in subsequent years as well. By the 1960s, the government started various programs aimed to boost female participation in mechanized agricultural production, as well as to find the main reasons for the low number of women working with new technology. New incentives were introduced as well. Thus, a resolution singed by the Soviet of Ministers on January 14, 1969, provided additional days off to female mechanics (increasing their paid leave from 6 to 12 days) and lowered their production norms by 10 percent. Collective farm chairmen were also required to give women the newest and best machines that they had, mostly to ensure that women had the most comfortable seats and the pressure relief systems for easier operation. Furthermore, in 1975, the government lowered the retirement age for female mechanics by 5 years, thus providing state pensions at 50.[17] Nonetheless, the skilled jobs of mechanics remained largely in men's hands, mostly because of the physical hardships of these jobs and a persistent negative attitude in the entire village towards women who spent most of their time away from home in men's company.

Propaganda campaigns were also formulated to recruit more women to become mechanics. Young and successful women were readily promoted to brigade leaders, and some were rewarded with medals and elected as a deputy in the Supreme Soviet of the USSR.[18] Entire families and dynasties of women mechanics were encouraged to send letters to Moscow-based journals and magazines to share their words of encouragement for aspiring mechanics. One of such letter read:

> Ever since I was in school, I envied in a good way those women who did grain harvesting with the use of combines. Our father is a tractor driver. And two of my sisters, Galia and Lena, graduated from a technical vocational school and became mechanics. And I also want to join the family ranks by becoming a mechanic in the future![19]

In 1973, the government established a prize named after Pasha Angelina, which was awarded to the best-performing woman mechanic. The names of the awardees were publicised in massive propaganda campaigns to encourage female participation. Z.A. Gribanova, M.A. Gureeva, A.S. Loginova, V.V. Mordakina and dozens of other women became Heroes of Labor and national celebrities for their achievements in the agricultural sector.[20] Women mechanics also established a club named after Pasha Angelina. They went to the club to discuss their labor conditions, hardships and challenges of working the machines, as well as to exchange some tips for making their lives easier. Although women who went to vocational schools to become mechanics were often teased for their effort, their successes and achievements often convinced even skeptics of the benefits of using skilled female labor in the countryside.[21]

Women, when they shared their stories of success, often emphasized that they were better qualified to do the job compared to men because women were more careful and attentive to detail than their male counterparts. P. Shabalina shared with *Krestianka* a typical life story:

> I started to work on the combine a long time ago during the war when we, the women, had to do the work of the men who left to fight at the war front. Recently I turned forty-eight. But I am still a good match to the young ones when it comes to work. Here, during the harvesting seasons, the female qualities like frugality and attention to detail bring

real results. I don't let a single grain go to waste. And the machines function longer under my care. My [tractor] did not break down even once in 4 years of service. And all thanks to maintenance, and nothing else! I try to work carefully, precisely. This is my only secret of success![22]

But the main incentives for women to become mechanics came from the chairmen of collective farms. If they were interested in recruiting women as highly skilled mechanics, they offered numerous financial rewards. As a consequence, the incomes varied greatly from farm to farm, as they were largely dependent on prizes for excellent performance. This meant that in some locations women mechanics could earn high salaries. One chairman, for example, sent several young women to a vocational school and later entrusted them to work a massive plot of land for corn cultivation. These young women did well on their plot, were rewarded for their efforts, and eventually formed two permanent female teams of mechanics.[23] In another example, in 1977 in the Millerovo district of Rostov region there were 30 female teams that employed over 300 women mechanics. Their motto was, "if it is done by women, it is done well!"[24]

Day and vocational schools in the countryside were also a crucial component of the recruitment efforts in the 1970s and beyond. New programs for training women to work the machines were introduced to regular high schools in the countryside in 1969, and the staffing of these schools and the range of programs improved substantially in the 1970s.[25] As of 1975, such programs were available in almost 10,000 rural schools nationwide, and the number increased to 22,000 in 1980 and 27,000 in 1985.[26] But not all graduates who finished these schools aspired to apply their skills to practice and make machines their profession. As many as 67 percent of high school graduates in the countryside wanted to go to college and continue their education. The most popular profession was that of an engineer, teachers came second in their popularity, and doctors were third. Even among more agriculture-oriented professions, the most popular were the jobs of agronomist and veterinarian.[27] Another option for training included vocational schools that emphasized on-the-job training while making general education courses less rigid than at college-prep schools. Thousands of mechanics graduated from vocational schools, and in the 1970s and 1980s, one-third of students in these schools were young women.[28]

In 1981, Soviet agricultural sector employed 23,000 women mechanics, although the number had declined to 17,000 by 1987.[29] But the working conditions remained challenging and even harmful. The main problems cited were related to: poor microclimate (40 percent); air pollution (30 percent); excessive exposure to exhaust gases (40 percent); excessive exposure to chemicals (12 percent); skin problems resulting from contact with oils and liquids used in machine maintenance (11 percent); noises (24 percent); constant vibration (45 percent); and finally back pain (40 percent). Tractor drivers complained of pain in their arms and legs (70 percent) and in their entire body (28 percent). As a result, the majority of mechanics (65 percent) noted that they were prone to extreme exhaustion, fatigue, migraines, dizziness and a range of other ailments after an eight-hour work day.[30] Women routinely reported that "in the winter time we get many colds [because] our feet are often wet, and we cannot bend our backs after hours of vibration."[31]

A lack of heated garages and storage facilities also made working conditions unpleasant for those who did repairs on the machines. Often, repair shops did not exist at all, and women worked alongside men under the open skies. One woman recalled that she was forced to quit the job because her arthritis became intolerable after doing repairs in the cold and rain. "We do not have a heated garage," she recalled, "and repairs? We do not have a shop . . . and if the tractor breaks down, please, fix it yourself!"[32]

Extreme fatigue, physical demands of the job, vibration and overexposure to all weather conditions resulted in many chronic illnesses among female mechanics. According to statistics collected by medical professionals, a third of all women who worked on a tractor for between 1 and 3 years had chronic illness; half of women who worked the machines from 3 to 10 years had acquired health problems; and the rate of chronic illness went up to two-thirds for women who were employed in these jobs for more than 10 years.[33] A tractor driver, Godunova, wrote to a journal,

> I am curious to see how many former tractor drivers who are now pensioners we have in the country? I would guess not many, at least not from my collective farm. And why? Because they do not make it to the retirement age.[34]

In the 1970s, the government authorized the production of new tractors that would be suitable for female use in the hope of recruiting more women to do these mechanical jobs. New tractors had adjustable seats to allow women who were short to reach everything easily. Vibration was minimized, as well as noise levels. Women had to have priority access to the most modern combines and tractors with the highest specification, especially equipment with pressurized steering wheels. In 1978, a new line of tractors, combines and other agricultural machines were approved as suitable for female use. Yet collective farm chairmen did not take these provisions literally, often denying women priority when better machines were available, and used their complaints as an excuse to relegate women to other spheres of employment. Most chairmen, who were predominantly male, did not perceive mechanized jobs as suitable for women and had no problems transferring these women to other, less-skilled and less-mechanized jobs. N.I. Miroshnichenko spent 10 years as a mechanic and always outperformed production norms. Yet eventually, she was forced to quit by the negative attitude of her farm's chairman. "I thought that now the administration will understand our needs and that we have family and domestic chores in addition to our jobs," she complained. "We hoped for some help, better working conditions, and an aid to help us with the repairs." But nothing happened on her farm.[35]

Parents also rarely approved of their daughters' choices to become mechanics. In one of many examples, a group of young women graduated from a vocational school and applied to work as mechanics at their collective farms. The mothers of these young women came to the chairmen and retracted the employment applications submitted by their daughters.[36] The same went for husbands. A traditional attitude that condemned women's long absences from home and week-long stays in the field did not change over time. Tractor driver N.B. Pereverzeva proposed to organize family teams where husbands worked alongside their wives and thus could be constantly aware of what their spouses were doing. Her personal experience was successful, but the system was not implemented widely and had only a marginal impact for women at large.

Contrary to governmental orders, some collective farm chairmen intentionally assigned women to work the oldest machines that constantly broke down, needed repairs, and were not realiable enough to meet production quotas. They pursued this practice with the hope that women would not be able to withstand the pressure and deal with the repairs, and would quit on their own. However, most women learnt to carry out repairs as part of their training and jobs, and only sought to take better care of their own machines. "I do not entrust my tractor to anyone," wrote G.A. Koshuniaeva, "because only I can care for it as if it were a baby."[37] Another woman complained that once she convinced her parents to let her work as a mechanic and then got her training, the chairman laughed so hard at her that she was deeply offended by his attitude.[38] In yet another example, women were given such poorly running machines that even a skilled repairman had to give up. He further concluded about these women that they "better go home and bake some cakes than work here because these tractors are nothing but old shoes." As a consequence, only 4 out of 11 women mechanics kept their jobs.[39] Furthermore, uniforms were rarely available to women. Women complained that all uniforms were in sizes extra-large and bigger, and even those were not readily available to all women in sufficient quantity.[40] Only rarely collective farms treated women mechanics preferentially, although these cases are important to note. A collective farm Suvorovskii authorized a local tailor to create uniforms for all women on an individual basis, ensuring that each uniform fitted perfectly.[41]

In the 1980s, women continued to claim their niche in this occupation, probably with a greater rate of success than ever before. In the Millerov district of Rostov region a quarter of all mechanics, or 340 people, were women, and the entire region employed almost 2,000 female mechanics in total. Their children received priority placement in childcare centers and priority access to specialized stores with deficit consumer goods.[42] They received uniforms and were provided with public transportation to and from their workplace. Their workday was divided into two shifts, thus allowing women to work fewer hours and have vacation days even in high season.[43]

The demise of the Soviet Union was detrimental in the short-term to agricultural production as it ended many state subsidies to this sector. It also exposed significant shortages in equipment, and even when available, this equipment was often run-down, old, and outdated. Approximately 70 percent of all machines had to be replaced, yet at the same time, the supply of new tractors declined 15 times compared to the pre-reform era; of combines, 21 times; and for other machines the decrease in the supply was anywhere from 13 to 54 times overall.[44]

In 2005, the government listed mechanization of the agricultural sector as one of its top priorities and authorized substantial subsidies to finance purchases of agricultural equipment. Thanks to preferential financing and state-subsidized loans, in 2008 farmers purchased over 23,000 tractors and 12,000 combines. This was the highest rate of buying new equipment ever since the demise of the Soviet Union in 1991. Although female participation in mechanized agricultural production declined, women nonetheless continue to play a significant role. They might be few in numbers, but they strive to be the best in what they do, or as one of them put it, they have "all the intentions of keeping the leadership position in our hands."[45]

4　Women at the animal wards

Success is a ladder you cannot climb with your hands in your pockets.

Unlike mechanized sectors of the agriculture, manual work at animal wards had always been considered a woman's task. Even before the coming of the Soviet power, women had been charged with and solely responsible for taking care of any domestic animals that the family owned. All jobs in the collective farm system that involved livestock remained exclusively feminized and were typically low-paid.

Starting in the early 1930s, animal wards employed women in large numbers, and women constituted 96 percent of the labor force. Most work was in caring for and milking cows, a job that was more labor-intensive than most other jobs done in the countryside. Unlike most other tasks, women who worked with livestock were employed year round without any breaks because they never had a low season. By the late 1930s, between 750,000 and 1 million women worked with cows.[1] These women were charged with all aspects of the process, including receiving and caring for newborn calves; growing grasses and collecting hay; bringing water from local wells or water sources for drinking and for cleaning barns and animals; and collecting manure and spreading it on the fields.[2]

World War II disrupted the typical production cycle as it took away many men to the front lines. Although the war years were challenging and demanding, the villagers were all too eager to revive their animal wards at the end of the war. Collective farms experienced a remarkable recovery in the first post-war decade, mostly thanks to the labor invested by peasant women. Many women who worked in animal wards were rewarded as overproducers in the country. Vera Rybachek became the superstar of the Khrushchev era and was the leader's favorite overproducer when it came to milking cows. Vera's portrait was published on covers of most Soviet-era journals and magazines, and she had two Medals of Lenin and the Golden Start of the Hero of the Socialist Labor by the time she was 27 years old. In 1958, Vera was recognized as the best milker in the nation. She learned to manually extract 16 liters of milk from a single cow in less than three and a

Figure 4.1 Milkers from a collective farm "Novoe" are working in a local pasture, in 1950, Russia, Vologda region. The farm was rewarded three times with the Red Banner, for record milk production.

Source: Photograph taken by Iarin. © Courtesy of the Russian State Documentary Film and Photo Archive, Krasnogorsk, Russia

half minutes, which worked out at 350 liters of milk a day or 56,000 annually, out of an animal ward of 18 cows with which she was entrusted. Even though she occasionally complained of her working conditions, Vera nonetheless remained loyal to her lifelong passion of caring for cows.[3] Vera spent most of her life as a widow with eight children, and she remarried at age 75 to a 94-year-old man. As a wedding present, she asked for . . . a cow![4] Such over-producers were envied by their neighbors, but nothing could dampen their love for animals.

The milker's workday revolved around three milking sessions. The first milking of the day was between 4am and 5am, and the last usually started around 9pm. Even though milking machines could have reduced the labor needed, they were rarely available, and women spent their lifetimes waiting for them to arrive to their collective farms. "We work in sub-human conditions," wrote one milker.

> We go around in dirt up to our ears because [we get no help], and we need to wear rubber boots both in the winter and summer time. There is too much draft and too little insulation in the wards, and the drinking vessels freeze up and we cannot keep up with defrosting them soon

enough. We stay at the ward until one in the morning to clean and care for our cows, and then we come back to work already by five in the morning.[5]

Because all milking was done manually, women were responsible for caring for 10 to 12 cows at a time, although some women oversaw as many as 18 cows. Just to provide enough water for her ward, a woman had to manually carry 80–100 buckets of water daily, a task that took on average three or at times up to five hours to do.[6] In the late 1950s, less than half of all collective farms with animal wards supplied running water to the wards, and manual milking was still the rule of the day. Less than 5 percent of all cows were milked by machines, and the rest of milking was done manually.[7] In addition, women spent up to two hours a day cleaning the manure, and they also had to bring the milk in containers to a centralized location on the collective farm. Milking itself was a repetitive process that occupied most of the day. A woman made some 140–300 moves a minute when milking, or 50,000 moves a day, and spent on average five hours a day doing just that.[8] Even in those rare cases when machines were available, milkers were expected to bend and lean over on average some 400 times a day to assure the proper functioning of the device.[9]

The situation was exacerbated by a negligent attitude of most chairmen to milkers' work. Most collective farm chairmen felt that women would not leave their cows no matter how poor the situation was, and hence they invested little thought, time and effort into making the working conditions of these women better. They routinely failed to repair leaking roofs and malfunctioning heating (if they had any at all) in the wards, even after women complained publicly in centralized newspapers about their poor working conditions.[10]

Because milkers were mostly women, they also had to negotiate and balance their schedules so as to allow some time for their household chores. Because three milking sessions a day made the workday long, most milkers tried to rush back to their houses in between milking sessions to catch up on their chores there. Or at times they got back home from work at night and started another shift of cooking, cleaning, and attending to their houses. Women in this occupation hardly rested or had time off.[11] "You come back from a day of work, exhausted, yet there is your own cow waiting to be taken care [of]," said L. Starostina, a milker from Kalinin region. "I come home, cry, take pity on myself, but then life proceeds in its turn," and the milker continued her workday at home.[12] As a rule, the milkers were recruited among middle-aged and elderly women who had no qualifications or skills to work elsewhere but also, as most chairmen believed, those who were already disappointed enough with their lives to stop complaining. Some milkers did complain to local and centralized newspapers, but most accepted their fate and never asked for anything better.[13]

Animal wards were typically in a horrible shape, and also they sank in dirt, quite literally. All facilities were directly built on dirt, without paved access or foundations. Hence the spring with its snow melting brought so much mud that in some places even a horse-drawn cart could not get close to the entrance, and containers of milk were first taken out through the window and then carried manually to a cleaner place. Barns were old and they had rotten beams, holes in the walls, leaking roofs, and hay to patch the holes and close windows instead of glass. Running water was never available, and in the summertime women brought water from a local pond or river and in the winter they had to melt snow to provide water for the cows. All utensils and containers had to be manually washed with cold water, and the same rotten bucket was used to give water to cows, then to wash them, and then to store milk. Milkers rarely, if ever, were seen by medical professionals, and it was common among milkers and their children to come to work sick, as well as to have ringworms and undulant fever (brucellosis). Needless to say, cows also often shared the same diseases, especially brucellosis, but milk they produced was not marked or discarded. Instead, it found its way to common milk containers.[14]

Milkers worked the longest hours compared to all other jobs, either in agriculture or industry.[15] Furthermore, industrial workers were paid an extra 30 percent of their wages if their workday was uneven, interrupted, or long (in excess of nine hours), but these conditions were not applied to milkers.[16] Until the mid-1960s, milkers were paid in labor days but the maximum number of labor days allowed was capped by the state.[17] When the payment system changed in the 1960s in favor of salaries and standardized monthly payments, milkers' financial situation improved, and correctly so. Theirs was the only year-long and round-the-clock job, and it remained the heaviest and harshest of all occupations in the countryside.

The government's program to mechanize all spheres of production was only marginally successful. Although with time the percentage of all animal wards with machines increased, in practice most of the new transporters or water lines were not operational. As a consequence, some 70–80 percent of all animal wards still relied on manual labor to do the job.[18] Nominally, by the late 1980s, 76 percent of pig wards and 67 percent of facilities for cows employed some form of mechanization.[19] Yet more often than not, a close inspection revealed that "everything is supposed to be mechanized but nothing works . . . and we do everything by hand."[20] In real numbers, approximately 200,000 women still milked cows manually in the late 1980s, and approximately 400,000 of women manually brought food and water to their animals.[21] And when milking machines became available, some women were hesitant to use them at first. Raisa Sekacheva was sent to get training in using these machines, and she admitted that "milkers looked at these with apprehension . . . I, too, was a bit fearful to use them at first; after all, it was a new business for us."[22]

Poor working conditions, hard manual labor and long hours made work at the animal wards less than appealing to women. Even back in the 1930s, a third of all milkers left the collective farm on an annual basis, and by 1980s, the turnover of labor force in this occupation was about 80 percent annually.[23]

The lack of human resources eventually forced some chairmen to authorize milking cows only twice a day (and not three times) and to created morning and evening shifts for milkers. Women responded to these changes with enthusiasm. "If I work the first shift," remembered Galia Merkusheva, "then I am home [on a break] by seven in the morning, just in time to serve breakfast and send kids off to school or daycare . . . then back to the wards until eleven [in the morning] when I give everything over to the next shifter."[24] Moreover, shift work also allowed milkers to have two days off a week, a luxury unknown until the 1980s. As a consequence of these reforms, the recruitment efforts were able to bring young women into this occupation.[25] The number of milkers under the age of 30 doubled on those farms that used two shifts for milkers.[26]

An opinion poll conducted in the 1980s demonstrated that female milkers who continued to work round the clock were highly dissatisfied with their labor conditions (75 percent) and the length of their workday (86 percent).[27] They worked long hours and spent on average an hour and a half getting to and from the wards. But like other women, they spend the bulk of their free time doing chores (for a total of 4 hours 40 minutes a day) and rarely slept more than five hours a day.[28] These women did not want their daughters to join the collective farm, and 70–80 percent of all milkers hoped to send their children to towns for education and better lives.[29] Young women with high school diplomas were almost expected to move on to something better, and those who stayed in animal wards were laughed at and told that "they graduated from a high school only to clean cow's tails."[30]

Neither the change to a double shift by some collective farms, nor propaganda campaigns endorsed by the state could fully remedy the abysmal situation with poor staffing of animal wards. "Let's turn the next Virgin Land!" shouted Komsomol (Youth Communist League) banners in these recruitment campaigns. In the early 1980s, about 440,000 young Komsomol activists went to the countryside to work there, including as milkers. But most of these young people were on the farm temporarily, as a part of a team that came to give a hand rather than stay. Selected women stayed, and many did so well that they received awards for their achievements.[31] Yet the majority came and left in a very short time, without ever aspiring to make agricultural work their lifelong occupation.

Working at the animal ward had its share of occupational hazards. Work-related injuries were common, with a third of all injuries inflicted by animals on their caretakers. Falls on a slippery and dirty floor of the barn came second at 17 percent of all injuries, and some 8 percent of injuries were caused by melting snow and heating up water.[32] Back pains and various problems of the musculoskeletal system were the most common, possibly because women spent 15 percent of their workday leaning and bending forward and 27 percent of their time sitting on their heels squatting.[33] Women also inhaled fumes all day long, and the concentration of ammonia in the air exceeded the normal maximum by many times.[34] Women were likely to develop medical conditions on or related to their hands and arms. Arthritis was commonplace, but also itchy rashes, blisters with oozing and crusting, inflammations, furunculosis, and even frostbite.[35]

Even the propaganda campaigns to encourage overproducing women could not mask various health problems associated with the job. Maria Kudakova, for example, had numerous awards and prizes for her achievements in caring for and milking cows. Maria always stated that she loved her job and her cows dearly and that she did not want to leave her job for anything. But at the same time she shared that "two autumns back my hands hurt badly; I even had to go to a special resort [for medical help]. My health improved, and I returned to the farm. But shortly thereafter the pain came back full strength."[36] Nonetheless, women who stayed in their jobs were emotionally attached to their animals, and even called them their "daughters."[37] This emotional attachment did not wear out with time. Elena Malikhonova remembered that she cried every single time when she reared a calf and then saw it off to another milking facility. "I know that they are not going to be slaughtered," she said. "But I cannot help it and cry out loud!" Two other women shared their observations about their cows, as if they were talking about humans:

> Each cow has its own unique personality. Some are frivolous and short-tempered . . . but they all know their names and their place in the barn. If one cow attempts to occupy a wrong place, it's trouble for her and us! And also our cows are friends to each other. They attend to each other, licking and cleaning their friends as if they were cats. And if their friend is gone, they would get upset and worry, constantly mooing to show their concern.[38]

The main challenge that animal wards face in the twenty-first century is a great discrepancy in prices for agricultural products and the costs of maintaining a farm. In 2004, a liter of diesel was 10 rubles, whereas a liter of milk sold for only 4.5 rubles (retail).[39] Moreover, the mechanization of this sector lags behind industrialized countries by some 40 years, and in over 80 percent of all cases, the power equipment that is available is too aged and barely functioning. The import of foodstuff is four to five times higher than export, and the state subsidies were cut down by a factor of 20 (20 times less) in the 1990s. Hence, it is unsurprising that of 30 million Russian citizens who lived below the poverty line in the 1990s, 25 million were rural residents.[40]

Various governmental programs adopted in 2005 aim to remedy this situation and offer incentives to people to start or retain their own businesses in this sector. Subsidies form the core of these reforms, especially in the form of preferential and interest-free loans to buy agricultural equipment. Some results were evident almost immediately; farmers who did not hesitate to get loans and partook in the incentives offered by the state started to produce twice as much milk as those who did not change their production mode.[41]

Moreover, some milkers were recognized in the Kremlin for their outstanding achievements, while others were offered an opportunity to travel to Western European farms to learn from their examples.[42] Yet all in all, although the government initiative is worthy of applause, it is still too premature to claim any definite or decisive positive results from these programs.

5 Women as collective farm leaders and agricultural specialists

Women in the collective farms are a major force![1]

Women in the countryside were indisputably an important workforce, yet they rarely worked their way up to become collective farm chairmen or highly skilled agricultural specialists. There was a significant opposition to promoting women to the highest ranks in the collective farm system, and although the propaganda campaigns for gender equality allowed some upward social mobility for women, their numbers remained small and marginal in the high-paid skilled labor force. Some women commanded such following and respect that they became de facto leaders of individual farms. Official chairmen in these cases were willing to recognize these women's authority and reputation among other peasants by allowing them to host meetings and lead agitation and propaganda campaigns.[2] Yet these women were rarely, if ever, promoted to official positions of power and prestige.

By the mid-1930s, women could be found in collective farm soviets as elected delegates to speak on behalf of all female workers, and women routinely represented the interests of the animal wards and milking facilities. At the same time, women constituted less than 1 percent of all collective farm chairmen, and they accounted for only 5 percent of accountants, 8 percent of deputies, and less than 10 percent of all agricultural specialists, such as agronomists and veterinarians.[3] Ironically, the heavy political repressions of the late 1930s created new opportunities for women who partook in the Stakhanovite movement or were deemed trustworthy in this politically tense environment. The way the repressions affected the upper echelons of the collective farm leadership were gender specific, because they were predominantly male. The repressions took away thousands of able-bodied male workers, and more women than ever were promoted to oversee and manage various departments of collective farms. In 1938, Kh. Moliakova became the chairman of her collective farm, and soon thereafter was appointed to head the Agricultural Department of the Kiliniskii region.[4] E. Borisova, who had been previously recognized for her record-breaking milking achievements, became the director of a state farm in Staro-Iuriev district. In 1939, 20 women were promoted to head their farms or other agricultural departments. These were the first women to rise to high ranks in the social hierarchy of Soviet agriculture.[5]

During World War II, women were once again the unexpected beneficiaries of the otherwise terrible situation in the country. Mostly due to the universal military conscription that took away men from their farms, women were routinely promoted to lead collective and state farms. As a consequence, by late 1944, nearly 15 percent of all farms were run by women, and more than half of all accountants, team leaders, and animal ward managers were women.[6] But as quickly as it came around, this trend was reversed in the post-war years. In 1956, less than 2 percent of all leadership positions were staffed by women, and this number remained unchanged for the rest of the Soviet era. Even in less prestige administrative positions, women made up no more than 20 percent of all employees, and mostly as administrative assistants to farm chairmen.[7]

The lack of women in the top-ranking positions was recognized at a Party level, beginning with Khrushchev's speech in 1961. As Khrushchev himself joked in his speech, "you better have binoculars if you want to find a woman" in one of these leadership positions.[8] But main obstacles remained at the local level; not only were men reluctant to promote women, but women themselves perceived leadership positions as suitable only for men.[9] A. Kornilova, who worked as a collective farm director for 20 years, said upon her retirement that collective and team leadership is "a man's business."[10] As late as 1990–91, less than 6 percent of all leadership positions in the countryside were staffed by women.[11]

The same tendency existed among various professionals of the Soviet agriculture. Social pressure, harsh living conditions and the reluctance to stay in the countryside after receiving a post-secondary education were all responsible for assuring minimal female employment in professional jobs. Indeed, educational opportunities were hardly to blame for a severe gender imbalance among agricultural professionals in the countryside. All educational institutions were equally available and accessible to men and women, and beginning in the 1930s, the enrollment of females in agricultural colleges grew continuously. The Soviet government required collective farms to send at least 25 percent of all women who qualified for post-secondary education to agricultural colleges.[12] Moreover, all farmers who received their education at such facilities were required to spend 3 years upon graduation working on collective or state farms in the area in which they were resident prior to carrying out the training. In cases, when the graduates resented or opposed such measures, they were charged the full cost of their education and a 6 percent annual interest rate on top.[13] Women received college-level education on a par with men, and hence educational opportunities were available to them as much as their male counterparts.[14] Ironically, by late 1950s educational levels among women in the countryside were higher than among men, even though women continued to work mostly unskilled jobs and earned on average much less than men.[15]

Even though women generally preferred teaching and medical colleges to agricultural institutions, nonetheless hundreds of thousands of women were annually trained to become agronomists, veterinarians, engineers, and accountants. Yet all these professionally trained women had to face the fact that chairmen rarely endorsed female professionalism, were not willing to hire these women, and were less willing to attend to their professional needs. Labor conditions as well as opposition from spouses and the village commune were substantial enough for women to seek other employment opportunities, or even to work in jobs that they could have done without going to college. Working and living conditions were so inhospitable, one woman veterinarian recalled, that "we never had any coal or any peat [for heating]. How were we supposed to work? There was only one choice left: forget my profession and work maybe at a local store."[16]

Professions that were required in agricultural production were also not popular with rural residents. Of those parents who aspired to have their children educated at a college or vocational school level, only a third thought that an agronomist or a veterinarian or similar professions were good choices for their children. The overwhelming majority, however, wanted their children to learn industrial skills and stay in cities, or become teachers and doctors.[17] Equally, only an insignificant number of school-aged teenagers wanted to learn skills that would be relevant in farming.[18] This trend is unsurprising

Figure 5.1 Women from a collective farm "Chervonnyi zamozhnik" are preparing barley for sowing in 1938, Ukraine, Kiev region.

Source: © Courtesy of the Russian State Documentary Film and Photo Archive, Krasnogorsk, Russia

considering the workload of such professionals. As a rule, the agronomists' workday started at 5:30am off-season and 4:00am during high season and, on average, they were overseeing production in four villages (each at least 3 or 4 kilometers away) for an average land plot of 2,400 hectares (approximately 6,000 acres). During high season, there were no breaks, no days off, and no vacations. One female agronomist remembered how fortunate she was to get four days off in a single year after the harvesting seasons was over![19]

Only with time the wages and working conditions began to improve. Crucially, the production techniques also shifted in favor of hiring more agricultural professionals to supervise and attend to various aspects of the agricultural production. Combined with the expanding educational opportunities, by the 1970s the proportion of women entering a skilled labor force was more substantial. In the 1970s, for example, women made up a third of all agronomists and veterinarians and veterinarian's aides.[20] Yet still only a quarter of all trained specialists stayed to work on the collective farms, while the rest found jobs in other sectors.[21]

The structural changes of the 1990s greatly undermined the position of women in the countryside. Large numbers of women were still graduating from post-secondary institutions and acquiring valuable skills for high-paying jobs in the agricultural sector. Yet those jobs were largely non-existent at this time, when collective farms were falling apart and new private farming was not yet mature enough to hire skilled labor. More than ever, parents aspired to send their children to city centers to receive education, however, with agriculture undergoing changes, the chance of finding employment in the countryside was often nil. Some parents from the Black-Soil belt in the southern parts of Russia wrote that their daughter "chose to become an accountant . . . but there is no job for her; there is no place to apply the knowledge because industry does not work and the collective farms are in decay. It's such pity that our youth cannot use their skills and knowledge and cannot apply it to a good end."[22]

Unemployment and the declining quality of life increased rural outer-migration both among the younger and middle-aged generations. Along with unskilled workers, many professional veterinarians, agronomists, economists and other professionals left the village to seek employment elsewhere, and most of them found jobs that had little relation to their training and skill. The countryside was experiencing a massive depopulation, and many villages lost their professionals altogether. Some women used this to their advantage, as they were the only ones left to manage local farms. Thus, eight women headed local administration in various parts of post-Soviet Russia, carrying the heavy burden of preserving their farms with little or no state support.[23] Yet, overall the practice of transferring power into women's hands was not welcome. According to research carried out in 2004, only one-third of all respondents to a public opinion poll thought that women should and did have equal chances to men when it came to getting a highly paid position of authority in the countryside, while two-thirds believed that men were and should be preferred for such positions.[24]

Governmental programs adopted in 2005 were aimed at increasing professional female employment opportunities in the countryside. For example, young professionals were given preferential treatment for housing distribution—16,200 young families in 2005 and 15,500 more in 2006 received subsidized (free) housing in the countryside thanks to the program. Production rates saw a steady increase after years of declining, and the economic growth of the agricultural sectors stood at 9 percent in 2008. Over 30 professionals from rural areas were recognized in Kremlin for their outstanding contribution to national economic growth, and some of them were women.[25] Symbolically, for the first time in the history of Russia, the Minister of Agriculture is also now a woman. Elena Borisovna Skrynnik is not only a capable leader, but a role model to many women who aspire to professionalism in all spheres of life, including the agricultural sector.

6   Rural intelligentsia

Known to all, trusted by all, helpful to all.

Although educational standards started to improve in the Russian countryside as early as the second half of the nineteenth century, nonetheless the bulk of the educational effort remained in the hands of parish priests who were able to offer little beyond rudimentary literacy training to their parish. Moreover, even this meager training was accessible to only a limited number of children and adults in the countryside. The rates of illiteracy remained high, as most educational reforms and new schools were only available to urban dwellers, and two-thirds of all peasants, or 88 percent of women, lacked even basic writing and reading skills.[1]

The Soviet government brought with it not only an effort to restructure and collectivize agriculture, but also a campaign to eliminate illiteracy in the countryside and to make schooling universally available to all Soviet people regardless of their place of residence. A Soviet Decree of 1918 set up mandatory literacy classes for all rural dwellers ages 8 to 50, and administrative and later criminal charges were implicated for those who opposed the training. Women were expected to attend these classes on a par with men, and they mastered writing phrases such as "we are not slaves, slaves are not us" under a scrupulous attention of their teachers. A later decree of 1925 made elementary education universal, and by 1926 the literacy classes had already started to bear some fruit.

The main challenge in the countryside was not in approving or resenting any Soviet decrees but in finding a sufficient number of teachers who were trained to offer these classes and available to work in remote parts of the new Soviet state. The Russian countryside lacked educational opportunities prior to the 1920s, so could not produce its own educators and had to rely on teachers from the city centers. However, these teachers, in addition to being reluctant to leave their urban lives, were simply not capable of offering instruction to everyone. There were too few of them for the task. In the Russian Federation alone, there was a need to hire 50,000 *more* teachers in the countryside in order to staff all schools that were created in the 1920s.[2]

Various training facilities and programs aimed to remedy this shortcoming, and short-term crash courses prepared new and aspiring teachers in a matter of a few months. In 1927, rural schools employed as many as 230,000 teachers, most of them for classes in basic literacy, and that number exceeded the number for all high-skilled agricultural professionals such as veterinarians and agronomists, of which there were 45,000. In the period of 1927–32, the pool of teachers doubled, reaching 450,000 in 1933 nationwide, of whom 280,000 were in the Russian Federation.[3]

Rural intelligentsia, and teachers especially, became of great concern for local Party officials. Such officials routinely reported that teachers were hesitant to carry out "sufficient agitation campaigns" to promote collectivization, and they were equally hesitant to get personally involved in dekulakization or agricultural production. The teachers never fell short in personal rigor when it came to schooling and education, but "they were reluctant to harm their relations with kulaks" and even staged organized protests against collectivization, and especially against the abuses of dekulakization.[4]

The Party, predictably, wanted to recruit future teachers from among workers and peasants. In the late 1920s, approximately a half of all students in pedagogical colleges were children of workers and peasants, and this number was perceived as unacceptably low by the standards of the Communist Party. The goal, imposed and enforced from above, was to enroll at least two-thirds of students from the peasant-worker background in an effort to minimize the participation of the so-called bourgeoisie in educating the nation.[5] In 1932, a quarter of all teachers were members of Komsomol or the Communist Party, and the proportion continued to grow.[6] The wages continued to increase, doubling in 1927–32, and teachers were offered preferential treatment when it came to housing and social services.

In the interwar years the problem of eliminating illiteracy was largely a rural phenomenon. If in the academic year of 1927–28 almost all children in towns and cities between ages of 8 and 11 were in schools, the same applied to only approximately 16 percent of rural youth.[7] Rural teachers had to face not only excessive teaching loads but a lack of resources and considerable opposition from parents. Most parents were reluctant to send their children to school, especially during harvesting times when extra pairs of hands were needed in agricultural production. Teachers painstakingly went from door to door to explain to parents the merits of educating their children. Although some parents remained unconvinced, by 1932, two-thirds of rural children were allowed and supported by their parents to advance even beyond rudimentary literacy classes into middle-school education.[8]

At this time, the Soviet government repeatedly considered introducing mandatory schooling through to the seventh grade, but such efforts and discussion were interrupted by World War II. Although education was halted in many parts of the Soviet Union, teachers put up their best efforts to carry on with the program, and they were often supported by the local population at large. By late 1944, education once again emerged at the top on the immediate agenda, and in various parts of the country schools were allowed to reclaim their buildings that had been used as hospitals and dormitories during the war.[9]

After the war, various governmental programs seemed to emphasize the need to recruit more teachers to work in the countryside. Teachers received preferential treatment when it came to the acquisition of livestock; they were granted special loans to start or improve their own houses. Some non-monetary, complimentary programs also facilitated teachers' work in their gardening plots to supplement their incomes. Collective farms were usually willing to help teachers turn the soil, buy seeds, and work their plots.[10] The attitude to teachers remained one of respect and even mild admiration, even in the harsh post-war years.

The educational standards promoted and legalized by the Soviet government also changed over time. If the initial goal was to eradicate illiteracy in the countryside, by the late 1950s efforts were being invested into making high school education compulsory. In 1959, secondary education became mandatory for all Soviet children, and the implementation of that reform was complete by the mid-1970s.

To meet the new and ever-growing demand for educators, vocational teacher-preparation colleges were either established anew or enlarged in order to enroll more students. From 1965 to 1970, teacher-prep colleges produced 350,000 new teachers who went to work to the countryside, in addition to 370,000 teachers who graduated from 4-year colleges and universities. Each subsequent decade produced substantially more new educators than the one before; from 1976 to 1980, 420,000 graduates of teacher-prep colleges and 730,000 graduates of universities joined the ranks of the profession, and the number went up again to 520,000 and 770,000 respectively in the period 1981–86.[11] Although the shortage of teachers became less pronounced, the disciplines in the most demand and the most remote regions still failed to recruit sufficient numbers of teachers, and children who finished those schools had no grade for disciplines that they were supposed to learn as part of the state program.[12] "We no longer aspire to have a teacher of any foreign language because it is hopeless," wrote parents from one such village. "And we still need more teachers of math and physics."[13] Parents from another village seconded the sentiments and explained that "it is the second year in a row that pupils in grades fourth through to eighth get no instruction in algebra, geometry, and physics. Why? We are told that there are no teachers. Help us!"[14]

This problem persisted even though, in the Soviet system, all teaching graduates were required to work in the place where they were "distributed" to and had no control over their place of employment in the first few, usually 2, years. Anywhere from 80 to 90 percent of all graduates with degrees suitable for teaching were sent to work in the countryside.[15] Yet many were not in a rush to leave their hometown, and a third of all degree recipients from universities and a fifth from teacher-training colleges never took up their placement.[16] Limited campaigns to enroll more teachers were just such—limited. Thus, anyone who agreed to work in remote parts of the country as a teacher was accepted to college without an entrance exam (while all education was free and students were paid stipends). This program saw 60,000 teachers graduating in the 1970s.[17] Collective farms were allowed to pay supplementary stipends to its own youngsters who went to study at a teacher college and came back to work at a local school.[18] These programs, however, failed to address the shortage of qualified staff effectively and comprehensively. It was also significant that in their freshmen year of college, over a half of all students in teachers' colleges believed that teaching was their true calling, but this number dropped by a factor of 2.5 by the time the same students reached their last year of college.[19]

Nonetheless, rural girls still often found the profession of teaching appealing to them. Commonly, teachers in the most remote regions belonged to families where teaching was the main occupation for generations. In one part of Irkutsk region, for example, the village school employed 18 dynasties of teachers.[20] Young graduates, especially young women, also tended to stay in villages where they met their life partners. They were also lured by the fact that most remote rural schools were small and required a level of a personal interaction with students unseen elsewhere. In a village called Istobniki, for example, the entire elementary school enrolled five students in grades one through three. Instead of a school building, the elementary school used a typical Russian *izba* (a countryside house), and the first task of the day was to bring in logs from outside, start a wood-burning stove, and boil some tea. Such an idyllic setting could be both the main attraction and the main challenge to young teachers. Some found the setting "perfect and charming,"[21] while others could not handle the demands of the job. At 19 years old, Lena Samkova, for example, complained that it was part of her job to prepare wood for the school to be used for heating, to clean the school, and even to feed her nine young students breakfast. Once the children left school for the day, Lena had "not a soul around, just an endless sea of snow. It was unreal to find anything to break the loneliness and the slowly-paced routine."[22] Although some young women enjoyed the solitude, the majority found it too hard to handle and preferred to remain in the towns where they attended their colleges and universities.

Teachers often performed other functions in the countryside. They were the source of all learning and contact with the outside world, as they were usually the ones to collect and bring state pension payments, newspapers, books, and even basic provisions to remote places. One teacher recalled that every day on her way to work, she packed 15 loaves of bread and some sugar and salt to bring to local dwellers on her way to school.[23] Finding a suitable partner in this limited setting with few people was not an easy task either, and 85 percent of all female teachers were not satisfied with their personal relationships.[24] Consistently, this profession was staffed by women, as their share in teaching and educating the young generation never fell below three-quarters of the entire workforce. Finally, teaching implied long work hours. Once the school day was over, teachers routinely spent up to five more hours a day grading homework and preparing classes for the next day. They also had to constantly review census information and report cases of teenagers being absent from school; they were expected and at times required to work in summer camps and after-school sports programs; and they routinely performed onerous manual tasks to assure the day-to-day survival of their schools. All of this extra work was carried out, despite the fact that teachers' wages remained low. Even though schools slightly increased the workload of teachers, to enable them to claim extra wages, the average income for a teacher remained 15 percent lower than other jobs.[25] It is unsurprising then to see that on average, as many as 15 percent of all teachers left their calling on an annual basis to seek employment elsewhere.[26]

The radical changes of the first post-Soviet decade affected rural schools as well. Yet whatever the fiscal and structural consequences of that change, teachers remained deeply respected professionals in the countryside. As one resident put it, "only in the countryside you can see without any formal lessons that the respect for elders and teachers is everywhere . . . a teacher and a doctor are both great people. Our lives truly depend on them."[27]

Indeed, in addition to teachers, doctors and various medical professionals were also deeply respected for the invaluable services they provided in the countryside. Although the availability of medical services and problems with medical facilities will be discussed at length in subsequent chapters, it ought to be pointed out that rural medical centers often only survived and functioned thanks to the efforts of individuals and the dedication of medical professionals. With chronic shortages of doctors, almost 80 percent of patients in the countryside were treated by nurses and nurses aides, who learned their profession so well that their diagnostic skills and prescribed courses of treatment were rarely of a lesser quality than that of better-trained doctors with university degrees.

As was the case with teachers, most nurses and other medical professionals in the countryside were women. Women were a dominant majority in all medical training facilities, from nursing schools to medical universities. But in the countryside, most hospitals had to rely on the poorly skilled, least-certified women to staff their facilities. Of over half a million doctors with university-level training, only 40,000 worked in rural areas. In thousands of rural medical facilities there was not a single doctor, and it was not unheard of to have a nurse or nurse's aide as the chief doctor and hospital leader.[28] The wages and living conditions for rural medical professionals attracted few new graduates, and only the most dedicated remained employed in remote areas after the mandatory 2-year employment service required upon graduation.

Workdays were also long and included a wide variety of chores. Typically, any rural doctor worked at the hospital between 8:00 am and noon and then had three hours of outpatient visits at the clinic. At 3:00 pm, doctors started their daily visits to patients who needed to be seen at home, for daily injections or urgent matters. On average, a single medical professional was responsible for overseeing over 1,000 people in two to three collective farms, and thus it took a lot of time and effort to carry out all home visits. Once home visits were completed for the day, a doctor had to meet quotas for collecting medicinal herbs, which amounted to 50 kilos of two or three types of herbs, supplemented with 50 kilos of birch and pine buds and rowan berries. All paperwork had to be completed in the doctor's spare time. At weekends and during harvest time, medical professionals had to "assist" the farm, sorting potatoes or stashing the hay. In spite of the lack of medical professionals, many urgent cases had to be attended around the clock when there was a need for medical services, especially in cases of childbirth, which often had to be assisted by the same staff member who had already worked a full day's shift.[29]

The economic problems of the post-Soviet era affected medical professionals in the countryside as much as anyone else, if not more. In the first post-Soviet decade, many medical facilities were in need of repairs and became severely dilapidated, and even the most urgent needs of medical professionals were not met. The lack of transportation made getting to patients a challenge. In remote regions, nurses had to walk several miles to get to their patients, most of the time in bitter cold or during heavy snows. Local residents recognized the effort it took for these professionals to get to patients' homes, and often put off seeing them until the health problem became truly urgent and produced significant complications. "People in remote villages are different [than urban residents]," wrote one female doctor. "They wait to the last moment without complaining much. They never call you at night [because] they fear causing too much trouble . . . Sometimes you ask them about when they got pain, and they say, well, last night." But, even when a doctor got frustrated over a prolonged wait, many patients thought it was wrong to bother a doctor at night.[30]

In 2006, the Russian government financed new training programs for aspiring medical professionals in the countryside, and attempted to stimulate professional activity and interest in service in the village by offering monetary compensations and financial rewards to rural medics. But these reforms are still largely in the making, and much of the medical care that is available in the countryside is the result of dedication and hard labor that these women share and perform out of their sense of obligation and love for their patients, and not for any immediate monetary compensations or rewards.

A village life also attracted a number of artists and other professionals dedicated to bringing culture to the countryside and serving the cultural needs of the rural population. Historically, the church functioned as the main meeting point and center of all rural activities. But the displacement of the Church and religious repressions left a vacuum in areas where the Church used to function as a social institution, and the state was all too eager to create secular alternatives to the church for meeting the intellectual and cultural needs of its people. The most immediate response was the establishment of the reading houses, or as they were known, reading huts. This was the center of rudimentary learning, and also the place where village residents could get access to newspapers, journals, magazines, and state-approved books. In a way, these reading houses were primitive and miniature versions of libraries, and their spread and popularity was impressive. In the interwar period, 40,000 reading huts were established in the nation, and although they lost their appeal and popularity after the 1940s, almost 6,000 of them were still operational in 1970.[31]

The social institution that became central and predominant in the countryside was a club, which functioned as a miniature version of the Palaces of Culture that existed in larger urban centers. The club aimed to cater to all sorts of interests, and routinely offered movie showings; lectures by locals and visiting guests; hobby clubs; social get-togethers; music lessons and various instructional circles; and other forms of entertainment. These clubs were normally filled with chess and checker sets, folk music instruments, pianos and accordions, and, by local standards, treasure-troves of books.

Access to, availability and quality of clubs varied greatly from region to region. Some clubs barely managed to survive (and many did not), and offered nothing in a way of organized interest circles or quality lessons. Others attracted professionally trained entertainment managers, theater directors, and artists. These people could and did breathe new life into a dying place. Although colleges for performing and applied arts graduated as many as 80,000

Figure 6.1 P.A. Malinina (front), a Hero of the Socialist Labor and the Head of the collective farm "XII October," during a chorus performance in 1965.

Source: © Courtesy of the Russian State Documentary Film and Photo Archive, Krasnogorsk, Russia

professionals annually, these young people, once again, rarely aspired to spend the rest of their lives in the countryside, and only 8 percent of all graduates were employed in rural clubs.[32] As a consequence, only 40 percent of all clubs had professionally trained employees, and these people earned some of the lowest wages in the nation.[33] Poor working conditions and low wages meant that by the 1980s, only 6 percent of all people listed as employees of rural clubs had post-secondary education and were properly trained for the job.

However, this did not mean that clubs lacked initiative or were a total failure. Individual enthusiasm and professionalism were often enough to put together and lead performance groups that even aspired to local fame. Local choruses, orchestras, and bands were popular outlets of talent for those who participated in them, and provided much-needed respite and entertainment for those who saw their performances. Competitions that were sponsored and endorsed by the state were often instrumental to allowing such groups to survive. Those with ambition, talent, and drive could compete with their neighbors, and some rural, non-professional bands and folk ensembles even achieved fame. The winning performances in state-endorsed competitions were aired on radio and broadcast on television, and those who were especially gifted went on tours all over the Soviet Union and even to Bulgaria, Poland, Italy, and Holland.[34]

The twenty-first century, with its obsession with technology, complicated rather than eased the life of rural cultural activists. To be effective, and recognized, remote rural clubs had to offer high-tech equipment and internet connections to their members. This task seemed impossible at the time of a massive economic demise of the immediate post-Soviet era, but in the new century dozens if not hundreds of rural libraries and clubs raised enough money to buy computers and pay for internet connections. Visitors of clubs in Chuvashiia or Mordovia, for example, can virtually stroll the halls of the Tretyakov Gallery or the Russian Museum at will, or to be more precise, at the push of a button. In a rural setting that lacks most modern entertainment facilities and opportunities, such initiatives and advancements are central to providing quality life to its people.[35]

But all of that said, it should to be added that a village does not divide people into classes, professions or occupations—people are separated into "insiders" and "outsiders," *we* and *they*. The openness and visibility of rural life that makes any privacy impossible tends to equate people of different ranks; people are judged by the commune, not by their words but their deeds. A professional woman can be accepted or rejected by the commune based only on her personal qualities, sincerity of her aspirations, and service to the community. The ability to reassure, to create a safe and stable environment in which everyone feels heard and taken care of, as understood by all villagers, is what can make or break a woman's life in that special world called the Russian village.

7   Migration to cities and the position of newcomers

I consider myself an urban dweller now . . . but at night I dream of my village; my home place does not want to let me go.

The forced modernization of Soviet agriculture led to an unprecedented wave of migration from rural locations to towns and cities. The entire fabric of Soviet life was developed around "the ideology of sacrifice for the sake of future generations."[1] The first wave of mass outer migration was a result of the prolonged and brutal process of collectivization. In the second half of the 1920s and into the 1930s, up to 23 million people, mostly men of working age, left the countryside to seek new jobs, education, and upward social mobility in the cities, as well to avoid repressions in the countryside and the heavy labor of a peasant. Another 5.5 million men were absent from villages in the 1930s on a more temporarily basis; most of them either worked seasonally in the cities, or were in the military, or undergoing training elsewhere. By 1937, almost a third of all men and 9 percent of all women in the age group of 16 to 59 migrated out of rural locations to find a new life in larger urban centers.[2]

The political terror that began in the 1920s and reached its peak by the mid- and late 1930s affected predominantly, although not exclusively, the male population of the countryside.[3] The accusations of "counter-revolutionary activities" placed these men beyond the protection of the law and stripped them of any right to seek grievances and justice in courts, which refused to review cases in these categories.[4] Local authorities were all too eager to promote the Party line and often exceeded even centralized quotas in their zeal to "serve the state." The fact of accelerating repressions and excessive personal zeal was duly reflected in the letters these local Party representatives sent to Moscow with pleas to allow them to make repressions harsher, to increase the length of imprisonment and hard labor for political victims, to be allowed to shoot more people, and exile more people overall.[5]

Dekulakization, which became a "straight murder," caused significant uproar among locals in cases where the property and possessions of kulaks were shared among Party activists on the spot, in front of fellow villagers and without even waiting to see the repressed leave their homes.[6] Acts of revenge, vandalism and hooliganism by relatives of the repressed were also numerous, including the distribution of anti-Soviet leaflets and agitation, "shooting at the portrait of the leaders," arson of collective and state farm property, and breaking and damaging other farms' tools and possessions. Typically, the village managed to write these acts of violence off as mere vandalism done by local teenagers, who did not fall under the law on counter-revolutionary activities and sabotage. As a response, the Chief Prosecutor of the Soviet Union, M. Pankratov, repeatedly requested from the government that the law be changed to prosecute for sabotage and other crimes starting at the age of 12.[7]

The overall number of victims of the Great Terror from October 1, 1936, to November 1, 1938, was 1,565,041 people, of which 668,305 were shot.[8] Of peasant dwellers and especially kulaks of earlier collectivization campaigns, many ended up in special settlements. But those victims who ended up in special settlements did not meekly accept their fate but continued to oppose the system by running away. Ironically, the escapes were facilitated by militia units and commandants who were supposed to guard the settlers, and at times the guards themselves ran away along with the former kulaks.[9] A research project conducted in the 1990s in numerous Russian villages demonstrated that approximately a third of all rural families were affected by the social engineering projects of the 1930s.[10]

By the end of 1939, the urban population of the Soviet Union increased by 27–28 million people. This number was greater than the increase of the 50 years preceding the October Revolution of 1917.[11] At first, the migrants were predominantly male. M. Zolotareva remembered:

> My father visited us rarely and always in secret. He worked as a construction worker far away from the village and risked being forced to work on the collective farm. Men left the village one after another, only to avoid the collective farm.[12]

Men were the first to go, but women formed the second wave of migration. The reunification of spouses often took years. Zolotareva went on to say that her father eventually decided to rent a house close to where he worked and asked her mother and children to come live with him. Her mother was 34, which was considered to be an advanced age for the countryside, and neither she nor the children had ever seen a train and were scared of traveling so far. They managed to buy tickets, although they overpaid, and the train conductor helped them to get on because the family was nearly paralyzed with fear. When they arrived to Moscow, the mother decided to send a telegram to her husband, but she was illiterate, and so were her children. The eldest daughter had learned the words "as if a prayer," but the clerk at the Post Office refused to send such a telegram. To complicate things further, the mother was sick with typhoid fever. They left the train station and traveled by a train car until they spotted their father in a crowd. The mother arrived without her passport, as most rural residents did not have passports at the time, although passports were required for travel. When militia came to check on the new family, they fined the mother twice for traveling and resettling without proper documentation. When the militiamen came for the third time, the mother still had no passport because she could not get one, and she was deported back home. That was the end of the unsuccessful attempt to have a happy family reunion in the city.[13]

Indeed, women were not in high demand in the cities when it came to employment. Women were physically less robust than men and, with children on their hands, rural women were seen as a burden and extra problems for this "gigantic factory of socialism." This was one of the contradictions of the Soviet system. On one hand, the Soviet government needed to keep peasants in the countryside to feed its population and to sponsor its huge industrialization projects. But on the other hand, the government needed new migrants from the countryside to work on these industrialization projects because their success depended on the labor of (former) peasants.[14] But by the end of the decade, most married men had managed to bring their families to live with them, even if it had taken years to do so. At the very least, statistically the proportion of female workers in cities increased from 24 percent of the entire urban labor force in 1928 to 40 percent in 1940. Much of this change was due to rural outer migration.[15]

World War II came and brought with it massive demographic changes, and the male population suffered greatly both in urban and rural areas, as men were sent to war and some never came back. Nonetheless, the outer migration continued; the village still provided men to the urban industrial labor force, as they left their familiar surroundings in their villages, even if at a slower rate than in the previous decade.

The second half of the 1940s and early 1950s finally allowed the villagers to welcome back the soldiers who survived the war and were now being demobilized. Nationwide, by September 1946 7 million people were demobilized, including 4.3 million who returned to live in the Russian Federation. Of these, 60 percent went back to live in the countryside. In 1945, the gender imbalance in Russia was 1.8 women for each man, but the gap narrowed to 1.5 women to each man by 1950.[16]

In the first 5 post-war years, the rural population of the Soviet Union decreased by between 1 and 1.5 million people annually. After 1948, when the demobilization was complete, the shortage of workers, as well as their outflow, became especially noticeable. The workforce of collective farms, not counting state farms, shrank by 3.7 million from 1948 to 1950. Undeniably, the spontaneous and uncontrollable outer migration of peasants to urban centers was a substantial factor in undermining the labor potential of the Soviet countryside.

However, organized and systematic resettlement campaigns that were endorsed—or even forced—by the state also played a crucial role in the overall pattern of rural migration. Prior to World War II, Soviet government resettled some peasants to Siberia and the Urals in order to repopulate these areas. Migration was also prompted by the industrial boom of the 1930s and the subsequent need for workers in various rapidly industrializing areas, as well as the hectic pace of collectivization with its dekulakization campaigns and mass kulak deportations. Thousands of other people became victims of purges and were forcefully relocated during various waves of political repressions simply for expressing their dissatisfaction with the General Line. But the victims of forceful relocation usually ended up in Siberia and the Far East, with few hopes for anything but mere survival. The post-World War II migrants were often lured by the financial benefits provided by the state, whereas the state aspired to repopulate newly acquired regions with an ethnically Russian population. Hence the main destinations included Sakhalin, Kaliningrad region, Karelia, as well as Saratov region, Crimean and the Grozny region, where many ethnic minorities resided prior to the war, and these areas became depopulated as a result of war-time ethnic forced resettlement to Siberia and Central Asia. The main "source" of new peasants was the central part of the Russian Federation.

Resettlement was done only by entire families and had to be strictly organized and systematic. Already in the first 5 post-war years, 160,000 rural families were resettled to new lands. Resettlers were treated preferentially (unlike in prior experiences) and were rewarded with special privileges such as small monetary subsidies and some household and personal items (usually a coat, a hat, a pair of felt boots, soap, and ropes). But crucially, the family received the right to a house with a gardening plot adjacent to it, all free of charge. Coming from regions where many rural families still lived in dug-outs, for many peasants this was the only option to improve their lives. They often ended up getting houses that had previously been occupied by ethnic Germans or Chechens, but they welcomed the opportunity, even if this took the family to a new and faraway place. Local farm chairmen were rarely glad to see people go because resettlement stripped their already understaffed farms of valuable human resources. These chairmen also had to fulfill the 'plan' that they received from the center, and any disobedience was persecuted as sabotage. Sometimes the number of families who were scheduled for resettlement proved truly astronomical, and extremely damaging to some regions. For example, the Kremlin thought that Kalinin region could easily, "without any harm," lose 15,000 families of working age, while Vladimir region could lose as many as 100,000 families. Orlov, Moscow, Kaluga, and Yaroslavl regions were all subject to resettlement quotas as well.[17]

In 1946–50, 156,000 families were resettled from the central regions of the Russian Federation. In the same time period, another 3 million people were recruited by the state to work on various industrial and construction projects as well as in the transportation industries. Needless to say, the departure of these people to work in factories was detrimental to many sectors of agriculture.[18] Thus, the Peremyshl'skii district of Kaluga region was one of the best producing in the nation, but once it lost most of its young and male workforce to state recruitment, it started to systematically lag behind in fulfilling the state plans.[19] In the Yuriev-Polsky district of Vladimir region, over 2,400 people, or 40 percent of its labor force, left collective farms in 1949–51. The remaining elderly men and women, teenagers and children could not keep up with the demands of agricultural production, especially because many had stayed behind because they were of poor health.[20] In 1945–47, Kostroma region lost over 73,000 people to recruitment, educational campaigns, and resettlement to Karelia and Sakhalin.[21] Some collective farms had no men left at all.[22]

State recruitment to work in various industries was often the only legalized chance for rural residents to leave their village and move on to a city life. Many young people were lured by prospects of this new and seemingly exciting life, although very few of them understood the hardships that they would have to face there. Back at home, most people resided in their own houses, however poor, and had some private space, whereas in cities they were assigned beds in dorms with communal facilities for all. Thousands of people were cramped under one roof, and hundreds of thousands of people lived in such dorms at any given time. The coal industry housed over 100,000 workers in dorms, and the building industry did the same for over 80,000 new recruits. On a personal level, individual stories were at times even more troublesome. For example, in 1952, the Construction Bureau of Chita region welcomed about 1,000 new workers. By the time they arrived, the dorm building was not even completed, and most rooms lacked windows and door frames, had unfinished walls, and had no heating. Workers were forced to sleep in their outerwear because it got so cold in the rooms. More than half of them were assigned a living space of 1.5 to 3 meters (or 13–18 sq feet). Unsurprisingly, "already after a few days some 300 workers decided to run away from the construction site. The director of the Bureau could not think of anything better than catching these young men at the train station."[23]

This story was by no means unique. The dorm of the Orsk Ministry of Construction housed 2,000 young women in a barrack that needed urgent repairs, had no glass in the windows, and had a leaking roof. The rooms lacked chairs, night stands, and tables. None of the rooms had radios or clocks. In the Kamyshinsk district, over 50 workers were housed in a storage facility for vegetables. In Stalingrad, the tractor factory did not have family accommodation, so married couples had to live separately, with husbands in male dorms and wives in female dorms, with no rights of visitation. A similar situation in a factory in Tulsk forced 36 families to live apart, and 114 families shared the same experiences in a Construction Bureau "Azovskii."[24]

These new recruits were also poorly provided with basic personal and household goods. In Vologda region, a store in a timber felling facility had no salt and no baked goods for months at a time. Bread was not always available, and at times was only supplied at three to five day intervals. The common eatery in Makarovsk timber felling facility, the only place where a worker could have a meal during the day, offered nothing but vodka, and meals were not cooked because the products were not available. Even when the food reached the chef, there was no salt for a month, so cooking and storing it was difficult. In another example, a common eatery had only five dishes for everyone, so each meal took up to five hours to feed the entire crew. In yet another timber felling facility, there was no water and the cook had to melt snow to use in cooking.[25] To complicate things even further, many newcomers did not receive their wages for months at a time, and they had no radios, books, magazines or any entertainment for after-work leisure time. On some occasions, hundreds of people did not go to work or chose to go back to their villages. And this is at the time when the Soviet law stipulated that any absence from a workplace or a change of employment at will without proper authorization was punishable according to the Criminal Code! The law was first signed on June 26, 1940, but remained in effect and was enforced until 1956.

According to official statistics, state recruitment was more detrimental to the rural workforce than voluntary migration out of villages for personal reasons. Thus, 768,600 collective farmers left to work in cities in 1949, and 80 percent of them were recruited by the state. Only 146,700 left villages on their own initiative and went to seek their fortunes in cities. Approximately the same ratio continued to dominate outer migration in 1950 and a few subsequent years as well.[26]

Already in the early 1950s, the Soviet government aimed to reform its program of peasant resettlement. The number of regions that qualified to release their workers was lowered; the new agenda included the resettlement of entire communities and brigades of farmers, and new incentives such as favorable taxation were introduced. But not everyone liked what they found in new places, and some returned back home to share their stories of misery and suffering. These returnees helped to create a negative image of resettlement and made recruitment efforts all the more problematic.

Khrushchev's Thaw affected the scale of this resettlement as well; the quotas for resettlers and new recruits were lowered to reflect a declining need to reinforce industries with new recruits. By the late 1950s, the resettlement campaign was largely over, evident from official data recording that in 1958, 90,000 people left the countryside voluntarily but only 1,600 of them left because of state-sponsored campaigns.

The city both welcomed and resented newcomers from the villages. Industry needed the unskilled labor that *derevenshchina* offered (to use a derogatory term applied to villagers which represented a sense of superiority of urbanites over peasants). Traditions and customs that villagers brought with them could hardly fit the demands and lifestyles of large cities, and the urban dwellers resented those people. As a consequence, the majority of migrants were marginalized and ghettoized, and failed to become integrated into the urban fabric of life. Yet these new workers rarely went back to their villages, mostly because very few of them had any place to return to and because they feared the reputation of being a loser who could not make it in the city. Even those former villagers who advanced rapidly on the social ladder and reached positions of importance in their workplace nonetheless were deemed to be known as *derevenshchina* by the rest of the urban community. A famous actress, Rimma Markova, whose stunning career was envied by many, reflected that "Muscovites consider the city their domains, and no one can claim a life there, especially if they come from a small provincial town or even worse, a village."[27] Former villagers were often at the very bottom of the social ladder, and shared the same notions of their worthlessness.[28] Yet they still accepted their new lives and could not even think of returning back from their "new home" to their villages.

These experiences almost created a trap for some, as parents did not want their children to endure the hard life of a collective farmworker and thus pushed their children to leave the village for the cities, usually to work there. Yet at the same time, most people could not adapt to the pace and the demands of the urban life—they worked long hours year round and had few, if any, options for social life and entertainment once the work was done. After the work shift was over, most workers were invited to attend lectures about proper communist work ethic or about increasing one's communist awareness and responsibility. These lectures were typically poorly attended, a fact that never failed to surprise local Party cell officials. Lectures on astronomy, medicine and love were much more popular with young workers, but were only offered sporadically. One Party boss in a factory in the Perm region complained that an evening dedicated to socialist work and the socialist work glory only attracted 90 out of 7,500 people who were employed by the factory, while "cookie cutter" events such as a movie showing were much more popular. Yet overall, the leisure time was poorly organized for new factory workers, and they often resorted to the most readily available form of fun— consumption of alcohol.

People who were fired from their jobs did not return to villages either but became "parasites" of the Soviet system and were repeatedly persecuted according to the Criminal Code. The increase in crime rate was directly linked to the new migration from the villages to cities, and the Chief Prosecutor reported that most crimes were committed by youngsters under the age of 25, and often by newcomers. Begging and homelessness were also registered among newcomers. Factory administration and local Party cells could not handle the inflow of new people and could not oversee them all on an individual basis. Ironically, some Komsomol leaders became the main instigators of drunken rages and at times "turned their rooms into a place of drunken orgies where one raid found over a hundred empty bottles of vodka."[29]

Nostalgia was a real problem for many newcomers as well. These people felt that their village was not just a place where they were born and raised, but rather was where they belonged. Victor Astafiev wrote, "I will not stop being a villager to the end of my days and will miss it always"[30] Those who came to urban centers knew that there was no way back and were ready to handle any hardships for the sake of staying in these cities. But they continued to feel homesick. When leaving her village to start a new life in Barnaul, Valentina Arinova believed that "you will rob yourself if you stay in the village." But as she experienced her new life working at a factory, she "came back in her thoughts over and over again to her small brook, to the smell of mushrooms in the woods . . . fresh birch leaves, the taste of blackberries and raspberries." And she started to question her decisions, to ask herself "maybe I was really mistaken. I am no longer part of my past but I do not belong to the factory life either. Who am I, a villager or a city dweller?" Valentina left the factory a few months later and, unlike many, decided to return to her roots.[31]

The Soviet Census of 1959 exposed a deficiency of males among the countryside population, especially over the age of 35. This deficit remained for at least two decades. As a consequence, the share of female labor in agricultural production was remarkably high.[32] In the age group of 40–49, 79 percent of farmers and working adults in the countryside were women, and the proportion was smaller only for those under the age of 20 and over the age of 60, mostly because women retired earlier than men.

The gender imbalance had other consequences as well. The proportion of married men in the overall male population in the countryside was higher than the proportion of married men in the overall urban male population. Therefore the number of available males was much lower in villages than cities. Even if all men aged 18 to 55 were married in villages, they would become husbands to only 50–60 percent of all women of marriageable age (the exact rate depended on the region). This implied that in the agricultural sector, about 12 million women aged 20 to 59, or over a third of all women of working age in the countryside, were not married and could therefore not count on their potential future husbands to help them physically, emotionally, or financially.[33]

In 1959–78, the city welcomed an average of 1.5 million villagers annually, and in 1979–88 the outflow of villagers continued at a pace of about 900,000 people a year. The majority of those who left their homes were young people seeking fortunes elsewhere. Until the 1970s, migration to cities was predominantly male. But later women outnumbered men in their resettlement patterns. Many rural parents saw no other choice but send their children, including daughters, to seek fortunes elsewhere.[34] According to official statistics, throughout the 1980s, 60 percent of all migrants were young women. A popular rhyme reflected this change, saying "the ships are passing and sailing by, only stupid girls count on their destiny [and stay in a village]."

The feminization of the outer migration from the countryside in the late 1970s and throughout the 1980s had negative consequences both for village and a city life. The gender imbalance within specific age groups was so pronounced that it was objectively challenging to find a potential spouse. But now the problem became reversed for young men who stayed in their villages. In 1989, for every thousand rural men in the age group 20–39, there were only 886 women in the same age group. As a consequence, the Census of 1989 registered that the number of men who never married was 1.6 times higher in the countryside compared to cities.[35] Some villages had no young women at all. One state farm in the Smolensk region reported that they had seven young males in 30 families but not a single woman of marriageable age. In another village, there were seven young males for no women at all.[36] In regions other than the Black-Soil belt, on average there were 79 women for 100 men, and the central parts of Russia needed over 300,000 women to compensate for this gender imbalance in the countryside.[37]

Letters to popular magazines routinely commented on this problem. A chairman of the collective farm in the Pskov region wrote:

> It is such pleasure to read about stories when young people, and especially young women, agree to go to the village and spend their lives there, because it is so hard for our fine lads to have no brides, and they are forced to remain bachelors even into their thirties.[38]

The newspaper *Komsomol'skaia Pravda* endorsed a propaganda campaign to convince young women to move to the countryside. They routinely published stories of farms that needed "women's hands," and never failed to present those farms in the best possible light. In 1987, 76 young women followed the call and moved to villages in the Kostroma region.[39]

Occasionally, such publicity backfired. A newspaper profiled two young men who worked at the state farm in the Chelyabinsk region and wanted to invite a young woman or two to work together and possibly marry. Instead of one or two, 25,000 young women wrote to these two men and expressed their interest in coming. Forty of these women did come to the village, and although most of them eventually returned home, eight stayed there permanently. The villagers in this case reflected that most women were unaccustomed to work in the countryside, had an overly romantic and inaccurate perception of rural life, and came to flirt rather than work.[40]

The proportion of women in agricultural production, which was high in the immediate post-war years, progressively declined in subsequent decades. Research showed that three-fifths of the decline among the rural population was due to outer migration of young women from the countryside. Leaving their home behind, women came back less often than men, who were occasionally attracted back to their villages by employment as mechanics with good pay and benefits.[41]

Many migrants who never came back to their villages still considered their return a possibility. In the 1980s, a third of all new migrants wanted to come back home, and more men wanted to return than women. Men cited the slow pace of life, better living conditions and a possibility to work the land as their main reasons for wanting to return home. Women, however, tended to criticize village for its lack of basic daily conveniences and poor social and educational opportunities. In 1986–88, 30,000 people went back to live and work on the farms.[42]

Another problem that affected the entire nation, beginning in the 1970s, was a low birth rate. Already in the early 1970s, the birth rate was barely at the level or lower than the death rate.[43] In the 1990s, the problem became even more pronounced when the death rate outpaced the birth rate for many years. As a consequence of the population decline, in the 30 years after the 1959 Census, the Soviet villages lost 10 percent of their population, and non-Black-Soil regions of the Russian Federation lost 42 percent of their people. In some of these regions the loss was greater than 50 percent.[44]

The deficit of human resources became evident in all spheres of rural life. In 1988, collective farms of the Kalinin region needed 40,000 more workers to staff its farms sufficiently.[45] In some villages of the same region the situation was so problematic that the campaign to "bind youth to the land" with incentives became meaningless, i.e. there was no youth left to "bind." There were many schools in villages where the entire school had only 5 or 10 pupils.[46] In 1988, in the Pskov region, arguably the most affected by depopulation, every fifth village (or 20 percent of villages) had no adults capable of work, and 300 villages had no permanent dwellers.[47] This situation made any efforts to return back challenging or even impossible for migrants.[48]

Previously, the government measures to keep the peasants on the farm had been harsh, yet effective. The Soviet government introduced a system of internal passports in the early 1930s, which were required for employment, travel, resettlement, and social benefits. However, rural dwellers were denied the right to have a passport, and when they needed to travel they had to apply to local authorities for one. However, the passports were rarely granted, and villagers reported that they were often turned away with the words, "your job here is good enough, go work on your collective farm."[49] Collective farm chairmen also gave teachers instructions to fail as many pupils in school as possible on the understanding that poor school performance would bind these youngsters to the land.[50] This system of discrimination was in place until the 1970s. But on the eve of having a new Constitution of "developed socialism," the Soviet government had to extend the right to movement to all its citizens. The Party issued a resolution in 1974, which stipulated that, starting in 1976, all citizens of the Soviet Union were entitled and required to have a passport upon reaching the age of 16.[51]

With few legal restrictions on movement, peasants' desire to leave the village became even more pronounced. Even though some people stayed behind, the generation of high school children in the 1980s came to consider leaving the village as the only viable option for them. Ninety percent of high school students insisted that they absolutely had to move to the city.[52] Many parents encouraged this idea or at least sympathized with it.[53] By the late 1980s, those rural residents who stayed behind often shared a common sentiment that they were somehow second-class citizens because they were "not good enough" for an urban life.[54] They resented the common attitude that ascribed them to a status of *derevenshchina* who meekly accepted what they had in a village without giving it a try in the city.[55] The Russian village was growing older and less populous as the majority decided to move on to an urban life. A lack of social services (sometimes because there was no one to serve), poor medical care, few possibilities for professional employment, few opportunities for social life and entertainment, and the aging environment of the village made rural revival a daunting task. In the disappearing villages on the non-Black-Soil belt, the young leave and the old die, leaving nothing but history in their place.

Part II
Private life

8   The politics of private life

The evolution and transformation of the Soviet Family Code

The October Revolution of 1917 brought many political and social changes to Russian society, and among them was the new attitude to the institution of family. The banner under which the newly formed Soviet government promoted various reforms was women's liberation from "kitchen slavery" and full equally with men, including equality in the domestic sphere. The goals of the new government were indeed grandiose: to take control of marriage and family life away from the Church, which previously registered births, marriages, and deaths and exercised a degree of indirect control over all aspects of family life; to simplify marriage and divorce procedures; to liberate women from domestic chores and childcare; and give equal rights and opportunities to all children regardless of their legitimacy.

Two important decrees marked the origins of this massive transformation. On December 18, 1917, a decree on "Civil Marriage, Children, and Keeping the Registry Books" established uniform requirements for registering marriage between members of various religious congregations without any concern for their religious background. It simultaneously recognized that only a civil marriage registered in specialized state institutions was a legitimate binding union. All religious ceremonies conducted after the issuance of this decree had no legal power and were not considered binding for the purposes of state regulations and laws, although Church-conducted marriage vows taken before the decree did not need to be reconfirmed and re-registered with the new authorities.[1]

The turn away from religious ceremonies and time-honored Church traditions allowed the state to simplify minimum requirements for registering marriage. Parental approval was no longer required for women over the age of 16 and men over the age of 18, and the only prohibitions against marriage were the close blood relations of the bride and groom, mental disorders, or the fact that either bride or groom was already married. Moreover, a new family was no longer expected to take the groom's last name but could opt for any combination of last names: last name of either a bride or a groom, or both, combined with a dash in-between. Prior to 1917, the bride always took the name of her husband as a symbolic act of submission and belonging; the name symbolized the unity of the family but also the fact that the new wife belonged to her new husband. Hence the change in this regulation was an important symbolic step in the propaganda of gender equality.[2] The same decree also stipulated that legitimate and illegitimate children had equal rights, and women could take delinquent fathers to court to establish their paternity and force them into acknowledging the child.

No less important was a decree on "Terminating the Marriage" issued in December 1917 in tandem with the one above.[3] It gave spouses complete freedom to decide when and why they wanted to divorce, and the new law abolished any need to specify the grounds for filing divorce papers, thus stipulating that explanations were unnecessary and irrelevant. The departure from the existing practices could hardly be more dramatic. Prior to 1917, only a Siberian exile of a spouse or an act of adultery proven in its entirety by two witnesses (with few insignificant additions) qualified as grounds for divorce. Needless to say, with such provisions the practice of divorce was minimal at best. In 1897, among the Orthodox people in the Russian Empire (who comprised roughly 70 percent of the population), there were only 1,132 cases of divorce. In the overall population, officially only 14 men and 21 women in every 10,000 ever divorced their spouse (although some arguably separated without registering the end of their marriage).[4] In 1913, the divorce rate was 0.015 percent of all married couples.[5]

The new Soviet government decided that since it was up to the spouses to terminate their marriage, it was also up to them to decide who got custody over children and how to divide any property the spouses had accumulated. If spouses reached a mutually satisfactory agreement, they could file their divorce papers out of court with the office of the registrars. Only in cases of disputes were the divorce proceedings taken to local courts. Inevitably, some gender biases (even if "reverse") made their way into this decree. Thus, if a wife had no means to support herself after a divorce, she was entitled to alimony payments that were set in proportion to her and her husband's financial situation. Yet the same did not apply to men who divorced their better-off wives; this provision survived in all subsequent Soviet laws on the establishment and termination of marriages. Yet despite its inherent flaws, these new decrees represented a drastic breakaway from pre-revolutionary practices.[6]

The first comprehensive Code of Laws on Marriage, Family Life, and Foster Care Rights and Obligations was adopted on October 22, 1918. It aimed to take into account a great diversity of potential family situations. Although it was more nuanced in its details, it did not depart from the two earlier decrees of 1917 and confirmed all regulations for marriage age, grounds for divorce, etc., provided in them. Women were also given equal rights to make decisions on where their family resided, whether it had to relocate for socio-economic or family reasons, and what last name the newly formed family and future children would have.[7]

But although the intentions were liberating, the attempts at a complete and immediate equally at times had negative consequences and threatened the very rights that the Soviet government aimed to protect. According to this new set of laws, women and men had equal rights to own property, and the fact of cohabitation, registered or otherwise, did not entitle either spouse to material possessions belonging to their partners. Although well-meant, such uniform equality failed to protect the rights of many women who worked their garden plots, took care of multiple children, and did all household chores yet had never been gainfully employed and relied on their husbands' work to make a living. These women were left unprotected and insecure after the marriage and often suffered miserably when their husbands, their husbands' incomes, and their husbands' material possessions were no longer available to them. The state simplistically decided that only those who earned wages could own possessions, and as a result women who spent their lives doing jobs other than wage labor were left penniless. Husbands had a right to take every-thing with them from their home, from furniture to the last spoon or pillow-case! The logic behind the law of "no common wealth" rested on the main ideological premise of the time that women who stayed home and were never engaged in public employment were the remnants of a bourgeois lifestyle and they had to be pushed, at times with horrible consequences, into the work-force. Yet in the new age of complete gender equality, as the state rhetoric went, every woman could and had to work and was capable of making a living and providing for her family on a par with any man. Especially hard hit were many women whose young children stayed with them after divorce, but who had to start their life anew as they lost all their household goods (although there were some limited provisions for alimony payments when women were in dire need of support). The overall sense of the new legal code was the complete freedom of private relations paired with harsh restrictions on propertied, material possessions.[8]

Similar complications arose out of further simplification of the divorce proceedings that this 1918 Law set up. If both spouses agreed to terminate their marriage, such termination was subsequently registered at the Registry of Civil Deeds (ZAGS, or vital records registry). If spouses could not come to an agreement on any aspect of the divorce process, including the divorce itself, then the case was taken to court and decided single-handedly by a judge with spouses present, or even in their absence. This system was quickly labeled a system of "a postcard divorce," i.e. send a note of divorce to court and get divorced without the mutual consent of both partners, or ever going to court, and it gave ample grounds for abuse. Often a woman learned of her husband's marriage to another woman and of their own divorce from a stranger well after it had been registered in the court, or at best the woman (at times men as well) merely received a postcard acknowledging that their divorce was complete. After centuries of marriages being written in stone, the new practice struck at the very core of the institution of family and gender relations as they existed in Orthodox Russia.

Section 132 of the Code also allowed parents to decide independently which of them would participate in providing financially for children, and to what extent. When both parties came to a common agreement, their arrangement was confirmed automatically during the divorce process. When no agreement was reached, the court decided on who was to pay child support and set a fixed sum to be paid regardless of real incomes or the basic costs of living. Children were also not entitled to the property of their parents, and parents, in turn, had no rights to claim anything that belonged to their children.

Section 133 of the Code gave full and equal rights to all children, even to those born out of wedlock. All fathers were obliged to participate in providing for a child regardless of whether the parents had been married.[9] Women had a right to force the father of her child into acknowledging his paternity as early as three months before the expected due date. Any man who denied fathering a child had two weeks to appeal to court and reject the claims of the mother. But a failure to appear in court (most commonly because prospective fathers did not even know about such claims) was considered a proof of acknowledging paternity. When a man took a woman to court to appeal against the claims of fathering a child, the court applied the presumption of accuracy to mothers' claims. "Evidence" that judges used to prove the paternity was based on statements such as "who knows better than the mother," "you need to trust the mother," or "someone has to be a father." In cases when it was undeniably proven that a mother cohabitated with several men at the time of conception and paternity could not be established, all men who had relations with this woman were obliged to pay a portion of child support and alimony payments, if needed. It was assumed that since every one of these men could potentially be a father of the child, they all had to pay for the probability of their fatherhood.

Historians have since argued that the resolutions and laws that favored children and protected childhood were dictated by the demands of the time. In post-revolutionary Russia, during and after the Russian Civil War, many children were left abandoned, orphaned, or with a single parent. The newly formed government, stripped of its already limited means by the demands of the Civil War and the post-war consolidation of power, had no resources to provide for children and worked hard to place the responsibility on the shoulders of its citizens, even when the parental connections were hardly probable or outright non-existent. Even when the child-support payments were undeniably misplaced or ethically questionable, the government justified its actions by pointing to the challenging socio-economic situation in the country.[10]

The shortcomings of the 1918 Code, especially in relation to child-support payments, became the focal point of discussion in 1923 when the Soviet government decided to rework the legal Code of 1918. Various drafts and proposals were rejected, and in 1925 the Soviet government decided that the discussion of the new legal code should be made public. Immediately, the questions of the legal meaning of a marriage (should it be official or de facto), of property rights of spouses, and child-support payments became the center stage of the discussion. The results of these debates were codified in the new Code of 1926 (KZoBSO).[11]

One of the main concerns was whether there was even a need to register a partnership between a man and a woman, and if there *was* a need, what the meaning of legal registration was. Although the socialist activists and Soviet lawmakers were influenced by proto-Marxist notions that a legally binding marriage was a dying institution, at the onset of the socialist change back in 1917, the choice to keep legal registration of marriages was dictated by the need to limit the functions of the Russian Orthodox Church (ROC). Immediately after the decree of 1917 that established only civil marriages as legally binding, the ROC announced that it did not recognize civil marriages. But the Soviet government retaliated in 1920 by closing down all consistories on charges that they undermined the functioning of the state and acted illegally by conducting religious ceremonies.[12] By 1926, the official rhetoric had it that "religious superstitions" such as marriage ceremonies conducted in the church finally died off among the Soviet people, and hence the question of legal registration was not a major concern of the state.

The realities of everyday life had proven otherwise. Because of various legal disagreements inherent in incomplete and confused decrees of the early Soviet state, by 1926 when the new code was under consideration, as many as 7 percent of all women in the Soviet Union found themselves in unregistered marriages with no rights to alimony payments and no protection in the face of the law. The overwhelming majority of these women were rural poor who either did not agree with or understand the need to register their marriage in the new offices of the Registrar, or had no means to do so.[13] Once their husbands left them, such rural women were left desperate. The liberalization of private life and the ease with which one could start or leave a family also created a massive rural movement of taking "seasonal" wives. Men cohabited with women during a harvesting season, but once the harvest, with its demand for extra labor, was over, so was their love and family life.[14]

In the end, the state was pressured by such massive destitution into reaching a compromise, which aimed to satisfy the real-life needs of women while still acknowledging the degree of freedom and equality that the Soviet official propaganda embraced. In the 1926 Code, the legal registration of marriages was kept intact, but common-law (de facto) marriages were also recognized and given *equal* status to registered marriages. The testimony of witnesses to the fact that a woman and a man cohabitated, had a common household, and raised children (if any) was sufficient to recognize a de facto marriage as legally binding and to claim full protection under the law. Although the intentions of such measures were aimed to protect all parties involved, and especially underprivileged women, as a result of the parallel existence of de facto and legal marriages some people found themselves in situations when both their de facto and registered marriages were recognized. However, there was no legal guidance as to how two marriages should be prioritized, which marriage took precedence, and who could claim what share of alimony or child-support payments.

The second major debate in 1926 was over the question of property rights. The new legal code abolished the largely failed system of separate ownership of material possessions in favor of mutual ownership. This change was especially important in the countryside where a large number of women still did not officially work; hence all possessions were purchased with incomes earned by men and thus all items of any value could potentially be taken away from women after divorce. Finally, the court no longer assigned child-support payments to multiple potential fathers. Women could apply for child support only after the birth of a child, and men had a right to dispute their paternity. Yet, as previously, in cases when fathering a child was a matter of debate, most courts were inclined to make decisions that favored women and children. Only when a man who was erroneously considered to be a father (but was not, biologically) could find the child's real father was he freed from his duties as the provider for a child.

The Code of 1926 also amended or supplemented some earlier provisions of 1917–23. To acknowledge the equality of sexes, the government decided to increase the minimum marriage age for women to 18 years, to make it the same for both sexes. All divorces were now settled in the Registry of Civil Deeds rather than courts, although a marriage could still be terminated in the absence of one spouse, and the spouse could then be informed by mail. In its basic structure, the Family Code that was adopted and signed into law in 1926 was not fully replaced, updated or rewritten until 1969, although its separate provisions (when taken individually) were modified significantly in 1936 and 1944.

Although it was never envisioned to replace the existing Code of 1926, the Abortion Ban of 1936 was a rapid turn away from well-articulated pro-choice practices that the Soviet state embraced prior to 1936. Previously, abortion clinics had a clearly recognized and well-established legal place in the hierarchy of medical facilities, and abortions were perceived as a matter of individual choice. However, the new code, as the name implied, banned nearly all abortions and made them illegal and anti-patriotic acts. To offset the expected shock of such a drastic reversal, the Soviet government paired this ban with maternal rewards such as child support for underprivileged mothers and mothers with many children; an increase in the number and accessibility of daycare centers for infants, toddlers and preschoolers; harsher punishments for failing to pay child support; and even some new limitations and revisions of the divorce practices.[15]

In the months preceding the issuance of this new law, the question on the ban was raised in press and in women's organizations with the intentions of receiving some feedback about abortion rights. After all, for many Soviet residents abortion was the only form of birth control available to them, no matter how unethical the practice might be. As it turned out, many rural women passionately supported the abortion ban and chose to concentrate on promised benefits and the expansion of the social-support system rather than any birth-control debates. They argued in favor of more extensive maternity and childcare leaves, better subsidies for families with multiple children, and even greater availability of childcare centers than included in the original proposal. In response to such requests, the government decided to lower the number of children required to qualify a family for special state subsidies from 10 or more children, as was originally intended, to seven or more children per family.[16]

The most popular provisions were those that included better medical services and medical care for expectant mothers and children. Yet, as was commonly the case with many Soviet-era measures, the decrees and reality departed drastically. The Abortion Ban stipulated wide availability of medical facilities in all villages, yet it never addressed the question of finding sufficient funds to oversee a massive extension of medical services to the rural population of the Soviet Union. The reality of the situation was such that even regional centers lacked doctors and other medical personnel to staff already existing medical facilities, let alone new ones. Some medical professionals went to the countryside only because they were forced to do so, by various policies of the state; hence they lacked interest in working in a village and, as a consequence, they commonly failed to earn the respect of local rural women. Typical of the period, women of one village in the Luzhskoi district of Leningrad region demanded that "an unpolished, rude, poorly educated and just stupid village *baba* [here, degrading for a simple-minded mature woman] who passed herself for a doctor" in the region be replaced on charges that she failed to pay attention to sick children brought under her care.[17] Although officially 18 percent of all licensed doctors worked in the village by 1930, in practice less than 10 percent were employed in rural medical facilities, while almost 9 percent of all medical experts remained professionally unemployed or underemployed in urban centers. The decree stipulated that there had to be at least one medical facility specifically designated for women and children in each rural center that linked several villages together. Since there was such a shortage of workforce, various proposals were considered to hire women with no medical training from among local rural residents to staff these centers.[18]

Similar situations existed with regard to other provisions of the government for child and maternal care. With the onset of a new agricultural year (typically starting in early spring), women demanded that their younger children get full care and access to various playgrounds. Yet, as another typical case helps to illustrate, the local collective farm officials disregarded such demands. When women of the collective farm (or *kolkhoz*) "Kuibyshev" in Shiroko-Kamyshinsk district of Saratov region (*oblast*) asked for a playground and infant care, the *kolkhoz* chairman bluntly replied: "You managed to live without playgrounds and infant care before, you can do without it now." Two leaders of *kolkhozy* in Odessa region even used buildings designated as childcare centers to house *kokhoz* administration.[19] Likewise, local administration in the Stalingrad, Ivanovo-Voznesensk, Voronezh and Yaroslavl regions denied state subsidy payments to mothers with seven or more children on the grounds that such provisions never applied to rural women, only urban residents. Only the intervention of the Chief Prosecutor (*prokuror*) of the Russian Federation, V.A. Antonov-Ovseenko, remedied the situation.[20]

Only the question of child-support payments could rival the issue of medical care provisions in terms of the debate it provoked among peasant women. Nearly every woman—a young *kolkhoz* girl to an agrarian professional— supported the notion that both spouses had to be present *in person* in court at the time of divorce to safeguard women from being left without any provisions. Equally unanimous was the opinion that the fees for divorce proceedings had to increase in proportion to the number of marriages, with each subsequent divorce being more expensive than the previous one. Many women supported the view that, after divorce, men had to pay 25 percent of their income in child-support payments for one child, one-third of their income for two children, and a half of the income for three or more children, to be divided equally among children. All of these wishes were taken into consideration and made their way into the new legislation passed in 1936. Especially popular was a provision for 2-year imprisonment for those fathers who evaded child-support payments.[21] These measures were new to the 1936 Ban but they were widely embraced by rural women.

However, political, social, and consequentially demographic, processes that shook the Soviet Union in the 1920s and 1930s handicapped many of the measures adopted in various articles on the said law. Collectivization and dekulakization campaigns, political cleansings, famines of 1921 and 1932–33, urbanization and industrialization, and simply the quest for a better livelihood that drove many to areas behind the Urals, stripped rural Russia of 30 percent of its male population, while only 9 percent of women left the village.[22] The ban on abortion meant to remedy this situation by "producing" more children and more workforce, especially male. Yet this very shortage and the constant outflow of males also made it increasingly difficult to enforce some measures such as child-support payments. Nearly 40 percent of all decisions to collect child-support payments could not be enforced as fathers could not be located,[23] though nearly all appeals made by women to collect child support from men were granted, and it was always men who paid child support. But although these measures might have discriminated against men, they were welcomed by many women left outnumbered—and under-supported—in villages across the USSR.[24] Many men complained that their share in the financial well-being of their children was exceedingly high (at 50 percent on most occasions for rural men), but this provision continued to exist throughout the Soviet era.

On the other hand, some women and later many scholars of the era saw in these provisions a turn to pro-male policies. All of these "favorable" policies implied and stipulated that aside from money, fathers had no role in childcare, and the position of women was codified as that of its main providers. Moreover, it forced many women, especially during the times when political terror was raging through the country and when the Soviet Union entered World War II, to turn to underground and illegal abortions in less than sanitary conditions. Although precise statistics are not available, many of these abortions were lethal, due to complications, bleeding, and infections. For many contemporaries and scholars such notions of limiting a man's role to the public domain and a woman's to the domestic were later confirmed by the inauguration in 1944 of various awards and medals for women with multiple children, such as "Mother-Heroine," "Mother's Pride," and "Medal for Motherhood."[25]

The pro-natal policies of the Soviet state become even more pro-family with the coming of World War II. The fact that various awards for motherhood were established in July 1944 allowed scholars to see in them a practical and expected policy of a government stripped of its human resources during the war. Indeed, already in 1941 various revisions of the existing Family Code went into effect, which were almost universally aimed at increasing childbirth rates. Such changes included the provision that all single and childless urban dwellers had to pay 5 percent of their income in extra taxes because they had no children to support, and the amount was set at 100 rubles annually for rural residents. This regulation was refined in 1957 when age brackets were put into effect; all men aged 20 to 50 without children, and all women aged 20 to 45 without children were subject to this special childless tax.[26]

In 1944, seemingly contradictory to its pro-natal and pro-children policy, the Soviet government issued "An Amendment" to the 1926 Code, which stipulated that women who had children out of wedlock were no longer given a right to claim child-support payments or appeal to establish the paternity of a man, thus taking away privileges the government had granted women in prior years. The logic behind this amendment was to force all couples into a registered marriage to facilitate the collection of child-support payments, and this mode of thinking might help explain the seeming contradiction of taking rights away from children while embracing a pro-natal stand. The government did offer some state support to women with illegitimate children but the amount of financial aid to "illegal" children was smaller than for legitimate offspring. This change was indeed significant as it affected nearly 25 percent of all children born in the Soviet Union, as they were categorized as "illegitimate." Even when a father wanted to pay child support, his pleas to do so were officially denied. On June 24, 1944, the National Committee of Jurisprudence (Ministry of Jurisprudence, or *Narkomiust*) appealed to Georgy Malenkov and Viacheslav Molotov to allow the registration of fathers who personally appealed to enter their name on the child's birth certificate and volunteered to pay child support. The request was never granted.[27] Ironically, if spouses separated without officially divorcing and a woman had a child by a man other than her husband, than she was still entitled to claim child-support payments from her official husband, even when the fact of "impossibility of conception" was undeniable (as, for example, in a case of a long-distance separation when a husband and a wife could not even keep in touch).[28] On the other hand, women who failed to register their relationship were left hopeless; as an example, in one instance a man fathered four children but then left his unofficial wife and children without provision. In 1945 the Presidium of the Soviet Union issued a decree to allow men who wanted to enter their name in their illegitimate children's birth certificates to do so, *if* these men later married the mother of their children.[29]

Some women, however, were forced by their destitute situation to search for ways to get around this amendment, and this was especially true of many underprivileged rural women in the Soviet Union. One of such ways was to refer to article 42, section 3, which stipulated that if any person took a child into full custody as a foster parent, but later "returned" the child, that person(s) was subject to child support payments in cases when the biological mother of that child was deceased, handicapped, or did not have sufficient means to support a child. In 1950 the Supreme Court of the Soviet Union decided that this section could be applied to parents without a registered marriage but with a mutual agreement that a man had full custody of his own children. As a result, some women could, and in fact did, successfully argue that fathers of children born out of wedlock took these children in their custody as foster parents or provided most of childcare and later returned children to their mother, and hence these children were entitled to child-support payments. Since men who did not have a registered marriage with the mothers of their children were treated equal to any stranger, in cases when it was established that a man cared for a child or children for any period of time, he was subsequently forced to pay child support.[30] Of course, this applied only to women who were not able to provide for their children independently. Nevertheless, throughout the 1960s for example, when the use of this section was widely welcomed, nearly 10 percent of all child-support payments fell under this category.[31] Statistics for some regions—Armenian Soviet Socialist Republic (SSR), Byelorussian SSR, and select others—also showed that the more liberal application of this section of the law resulted in a decrease in numbers of women who received state subsidies for childcare.[32]

Numerous appeals to allow fathers to enter their names in birth certificates of illegitimate children had little immediate result even when these requests were voiced by government officials. During one meeting of the Ministry of Jurisprudence in May 1950, most members of the meeting explicitly demonstrated their support for a notion that

> in regard of birth certificates for a child, when a child is small, it does not matter to him; but when the child reaches a school age and starts school, he begins to understand that there is no father. And, to avoid such psychological traumas, we propose that a mother have a right to write in [in the birth certificate] anyone [she desires] just to let a child know that he has both a father and a mother.[33]

Yet, as it took 6 years after the law of 1944 to voice such concerns, it took 18 years to override the law and see the change. But such provisions affected about 17 percent of all children by the mid-1950s,[34] or nearly 10.6 million children who were born while the specific provisions for illegitimate children of the 1944 legal code were in effect.[35]

The law of 1944 was a drastic departure from the previous decades when both common cohabitation and registered marriage were equally recognized. Hence all those who cohabitated prior to 1944 were allowed to register their marriage and "legalize" their common children. The results were predictable; most couples wanted to register their relationship after 1944 and made such legalization the ultimate goal of all long-term relations.

Simultaneously, the amendment of 1944 made the divorce process increasingly complex. It was no longer sufficient to file divorce papers; now, at least in theory, the court had to establish grounds for divorce and could deny appeals to terminate a marriage. After an application for divorce specified the grounds for such action, this announcement was subsequently published in a local newspaper. At the same time, the court was granted a right to determine the amount of child support and to resolve single-handedly any custody issues.[36] Although commonly ignored in large cities, these provisions nonetheless found their way into small villages, where at first a potential divorce became subject to rumor in the entire village, and then the community could oppose the divorce and deny the right to terminate a marriage if "silly reasons" such as "incompatibility" or "a lack of emotional connection and love" were cited in the divorce application.[37] Only the existence of another

Table 8.1 Official marriages in the USSR, 1940–54, in thousands

Year	Overall	Urban	Rural
1940	1,082	538	544
1943[a]	347	175	172
1944[a] (January–June)	*265*	*139*	*126*
1944[a] (July–December)	*317*	*186*	*131*
1944[a] (Total)	582	325	257
1945	1,106	637	469
1946	2,100	1,099	1,001
1947	1,859	865	994
1948	1,917	908	1,009
1949	2,018	1,002	1,016
1950	2,080	1,059	1,021
1951	2,125	1,085	1,040
1952	1,934	986	948
1953	2,073	1,087	986
1954 (January–June)	*1,094*	*559*	*535*
1954 (July–December)	*1,146*	*588*	*558*
1954 (Total)	2,163	1,133	1,030

[a] Without occupied territories

Source: *Sovetskaia zhizn' 1945–1953.* (Moscow, 2003), pp. 695–6

family, especially when a man took a new wife rather than a woman a new husband, with another child in the new family, qualified as undisputable grounds for divorce in most rural settings.[38]

Overall, the 1944 legislation clearly aimed to consolidate and solidify the Soviet family, and many subsequent family laws and regulations worked to the same end. For example, the decree of February 15, 1947, forbade any marriages between citizens of the Soviet Union and foreigners.[39] But already in the 1940s pleas were made to perfect and refine the existing Code of 1926, which by now incorporated the Abortion Ban of 1936 and the Amendment of 1944. But revisions and massive redrafting of the law would not materialize until 1969. However, abortions were once again legalized in 1955, in an attempt to undermine the wide network of illegal abortionists and combat the high post-abortion mortality rate among women.[40]

Starting in the late 1940s, there was a continuous discussion of the need to turn the 1944 code into the first All-Soviet Code. Whereas previous regulations and decrees required that all laws passed by the Soviet Presidium be incorporated into the legal codes of the Soviet republics, there had not been a single all-Soviet family decree or law issued to date. Hence on December 17, 1948, the Soviet of Ministers of the USSR formed the State Committee to prepare the basic guidelines of the all-Soviet Marriage and Family Law. Ironically, of the nine members of the Committee only one—G.M. Sverdlov— was an expert in family law, while all the others were party officials with no background in this kind of work. The torch of preparation work was passed to the Judiciary Committee of the Soviet of Ministers of the USSR in 1956, which submitted the final draft in 1959 for adoption by the Soviet of the Ministers. Yet the members of the Soviet "found it reasonable to postpone any considerations of the draft for an indefinite period of time."[41] Among all articles proposed to be ratified into law, the article on illegitimate children caused the most controversy among the members of the Committee. Hence, in 1963 the Central Committee of the Communist Party of the Soviet Union allowed the publication of the new law proposals in the press for public discussion.

However, the proposal was not published the same or the following year. In December 1966, the Chair of the Legislative Committee of the Supreme Court, M.S. Solomentsev, suggested that the publication go ahead because of the need to ratify a new Family Code, and the Secretary of the said committee, S.M. Isliukov, finally reported to its members in October 1967 that

> at the present moment the first draft of the marriage and family law, after some revisions in the committee, has been presented before the Presidium of the Supreme Soviet of the USSR and sent to the respective state departments. There is an intention to publish it in press prior to ratification for public discussion so that it can be subsequently approved by the Supreme Soviet of the USSR.[42]

Finally, the first draft was published in various newspapers and journals on April 9, 1968. The response was impressive; although exact numbers are hard to come by, the committee responsible for drafting the new legal code received over 4,000 letters directly, and the newspaper *Izvestiia* received over 5,000 letters with recommendations for changes and amendments to the proposed decree. The debates about the first draft were lengthy, especially when it came to the question of child support. But most active among all respondents were rural women, who felt that they were deprived of many privileges common to urban residents prior to the 1960s, and who saw the new legal code as an effort by officials to ensure that benefits would finally reach them. Prior to the 1960s, de-Stalinization campaigns presumably started an era labeled as Khrushchev's Thaw, which, among other things, resulted in the massive expansion of social benefits and the social welfare sector. However, the change of policies was non-consequential for rural people, especially women. During Khrushchev's Thaw, the 1944 legal provisions that many rural women felt were discriminating against them and their children were not amended or changed. Moreover, the most radical measures adopted during the Khrushchev era, the so-called "universal" rights to own a passport and get pensions and social benefits—was anything but universal. The passport was the single most important document in the Soviet era. Without it, a citizen had no right to relocate because every Soviet person needed a residency stamp to find a job and live in any location, and the stamp could only be found in a passport. So by denying rural residents a right to own a passport, the Khrushchev Thaw denied them any possibility of social mobility and even a right to relocate. Arguably more important, a social welfare system of paid maternity leave and paid child sickness leave was not extended to rural women. Similarly, a social security system of pensions for elderly residents had been in place in some form since 1930,[43] and new categories of people who qualified for state support were added in 1936 and 1941 to include military personnel and some widows.[44] In 1956, another decree of universal state pensions extended old-age benefits to all categories of Soviet citizens. Yet these "universal" provisions applied to all Soviet people *except* rural residents. Considering that of all elderly rural residents, nearly 55 percent were widows while only 13 percent were widowers, "the Thaw" was anything *but* a thaw to these women in the Soviet countryside.[45] One woman wrote:

We spent our lives [working] for the *kolkhoz*, we lived through everything. Our men served the country [especially during the war] while we worked day and night. We were hungry but we turned the soil with shovels, used sleds in the wintertime and carts in the summer to manually transport manure to the fields. We ate only the skins of the oats left over after cleaning, [we] were paid nothing but did not leave our jobs. Back then, there were no tools like now. We used sickles to harvest the grain by hand; we saved every single ear of grain.

And it was all for nothing, this woman concluded, because none of the benefits were extended to rural people.[46] Only in 1964 were rural residents given rights equal to urban residents and paid pensions after reaching the age of 65 for men and 60 for women if they had worked cumulatively at least 25 and 20 years respectively over their lifetime.[47] Although the retirement age was initially higher for rural residents than for industrial workers or professionals, it was lowered by 5 years in 1968 to make it the same for all citizens of the USSR.[48]

These changes in the social support system prompted many rural women to treat the proposed Family Code as one of many simultaneous changes aimed at improving their lot. It is highly possible that this understanding stimulated the massive response among rural women to the publication of the first draft in 1968. For the first time rural women learned that they not only had obligations and duties, but also rights. The new draft emphasized such notions as "motherhood and childhood happiness" and "a rewarded motherhood." The draft opened with the statement that "Soviet women are offered all necessary basic social conditions for a happy motherhood which can be paired with progressively more active and more creative involvement in the industrial and socio-political life [of the country]." A special third article announced "an equality of a man and a woman in all family relations," although this statement was contradicted by the subsequent section that talked of "universal protection of the interests of mothers and children" (not parents, but only mothers) and also by another section of the same article which detailed what this protection implied for "women-mothers."[49]

So when the new Marriage and Family Code was adopted on June 30, 1969, it included many of the provisions that initially sparked public debate. According to the new code, only registered marriages led to legal rights and responsibilities, whereas a de facto marriage (i.e. cohabitation) did not entail any legal status. When there were no children involved, spouses could divorce in the Registry of Civil Deeds, and only when there were property disputes or under-aged children were the divorce cases heard in court. The grounds for divorce were only "the irreparable dissolution of the family." The paternity question was decided on a voluntary basis. Any man could claim to be, and subsequently registered as, the father of a child, and common cohabitation was sufficient for women to take their partners to court in cases when they fathered children but denied their paternity. All children were given a right to have their father's name entered on their birth certificates regardless of their parents' legal registration.[50]

In the following decades, the Soviet Constitution of 1977 incorporated some of the rights and provisions previously stated in the 1969 Family Code. The Supreme Soviet approved the notion that every citizen's primary responsibility was to be a good worker and a good family person. Hence, family questions assumed center stage as an equal part to all other discussions. At first it was proposed to open article 35 (which dealt explicitly with gender equality) with the words "a woman has rights equal to those of a man." But such phrasing was criticized as it seemed to imply that men's rights were taken as a model, and women were allowed to share into those rights. Hence article 35 opened by stating:

> Men and women have equal rights in the USSR. The realization of these rights is satisfied by giving women rights equal to men's: in gaining access to education and professional training; at work; in work-related compensations and promotions; in public and cultural activities; as well as special measures that safeguard women's labor and health and special provisions that allow women to combine employment with motherhood. [Women are also given] legal rights, financial assistance and moral support to the fact of her motherhood, [including] protection of the childhood, paid maternity leaves and childcare leaves of absence, and a gradual reduction in the work hours per week for women with small children.[51]

Yet those who scrutinized the Constitution noted that article 53, which addressed family rights, seemed to contradict some provisions of article 35. Article 53 included statements that "in family relations women and men have equal personal rights as well as property rights."[52] Yet concern was immediately voiced that article 35 offered many rights to women, primarily related to her child-bearing ability, that were not extended to men. To achieve full gender equality, the lawmakers had to abandon the stereotype that a woman's primary function was to become a mother, whereas men could not have true equality when it came to domestic questions and especially childcare. None of these stereotypes were overcome at the time, and it was understood that

> the equality of men and women de jure had never been perceived and is not seen as the equality of their legal status. Simply equating men and women does not guarantee equality to women who, in addition to all those functions performed by men, also perform functions specific to her alone, those of the motherhood. Hence it follows that the true equality of men and women is possible only when women, in addition to having all the same rights as men, are also given special rights and privileges.[53]

Such contradictions, although seemingly extending many additional rights to women, also downplayed many hidden rights offered to men. For example, according to section 14 of article 53, men could not initiate divorce proceedings if their wives were expecting or if any of the children were under the age of one, thus making a point of protecting women. Yet at the same time, a right to a paid leave of absence in cases of child's sickness was only extended to women and never to men. Some argued that this regulation looked after women, but in practice it implied that only women were away from their workplaces for family reasons, while men were "good workers" who deserved promotions for their commitment to the workplace. As a result, it became nearly impossible to assure gender equality in employment, especially when it came to promotions and raises. Likewise, the practice of leaving children with their mother in 99 percent of all divorce cases seemed to benefit mothers, yet it took away from the roles—and responsibilities—of men in the domestic sphere while simultaneously freeing these men for career advancements and leisure.[54]

When in 1979 the UN General Assembly adopted the Convention on the Elimination of All Forms of Discrimination against Women, it offered a comprehensive example of how to provide special rights for women in their roles as mothers, while not impeding their rights as individuals (e.g. equal access to employment, equal pay, voting rights, etc.). Yet the Soviet government proudly announced in the same year that "in our country all measures of this Convention have long since existed in full . . . The Soviet legal code represents all provisions completely, and at times even exceeds the established norms."[55] Nonetheless, the Soviet Union ratified the Convention in 1981, thus bringing international gender equality norms to the Soviet Union.[56]

The coming of *perestroika* and the declining birth rate in the second half of the 1980s once again brought attention to "a woman's question" and the need to solve problems for many women, especially in the countryside. Mikhail Gorbachev pronounced that the solution should be ideological, meaning it should come from the realization that women had a right to, and should, leave their workplace and dedicate their lives to the no-less-honorable and no-less-challenging task of motherhood. Most official documents of the era began to talk about "the return of women to their true calling."[57] Although such a call was equally applicable to both urban and rural women, the government noted that there existed "an especially challenging and crisis-like demographic and socio-economic situation in the countryside."[58] As a result, the government adopted several measures aimed specifically at the rural population. The state embraced the notion that it was easy to "return" rural women to their roles as mothers of multiple children if these women were given extensive social benefits. Hence in 1990, the Decree on Rural Women stipulated an increase in the length of maternity leave for women, and in wages and benefits paid to mothers who chose to stay at home with children until children turned three. It also introduced a one-time payment to the mother at a child's birth, offered free food to expectant mothers, and issued a strict prohibition against the employment of pregnant women in jobs that required contact with livestock or heavy labor, from the moment the pregnancy was confirmed.[59] In terms of employment patterns and workplace equality, the Labor Code of the Russian Federation, signed into effect on February 1, 2002, confirmed most of the work-related provisions of the earlier decrees of the Gorbachev era and added further benefits. Thus, both men and women gained the right to take leave in cases when a child was sick, although on most occasions women took the lead in taking such leave.

At the time of the massive inflation of the 1990s, revision of the policy on child support payments was also required, and was achieved in 1994. Furthermore, a new Family Legal Code was signed into law in 1995 that aimed to solve some of the dilemmas of the inefficient and out-of-date 1969 Code, while simultaneously reconfirming other provisions. Only registered marriages were treated as legally binding, and common cohabitation, no matter how prolonged or how many children it produced, resulted in no legal rights or obligations. The 1995 Code further confirmed the right to seek divorce out of court administratively at the Registry of Civil Deeds with the mutual agreement of both parties, and when no children were involved. Spouses were no longer obliged to give grounds for divorce, as the state concluded that any explanation could be perceived as an invasion of privacy. If one spouse objected to divorce, then the court had a right to postpone the hearing for three months to give the couple time to reconcile their differences. All the decisions with regard to children were based on notions that all child-related provisions had to protect individual rights of the child.[60] The improvements in medical science also allowed for establishing paternity quickly and undeniably. Moreover, according to the 1995 provision, if a man desired so, his name could be written into a child's birth certificate as a father of a child without further inquiry into this matter, or a mother could take a man to court to ensure that he fully performed his duties and responsibilities as a father, most commonly by paying child support.

Yet however strict or lax the current legal code was, no law could claim unlimited and complete control of a person's life. Indeed, the life of rural women often followed a path that differed drastically from the provisions written into various legal codes. In some cases, women had more freedoms than the law offered, yet often the limits set by traditions and customs respected in the countryside were more binding and unquestionable than any official ideas about life and family dynamics in the Soviet countryside.

9 Marriages

I want a wedding like everyone else.

Until recently, historians of twentieth-century Russia tended to represent a rural woman's life as dark and backward prior to 1917, and as rapidly changing in favor of a more "enlightened" and free existence after 1917. This "before" and "after" periodization is partially justified by the dramatic revision of the family and marriage law after 1917. Yet despite such revision, there was more continuity than change in the real-life experiences of rural Russian women in the first half of the twentieth century; it would take the experiences and social transformations of World War II to drastically alter the fabric of rural life. One of the best examples of this continuity is the institution of marriage in the Soviet countryside. The common pattern of arranged marriages survived long after the establishment of the Soviet power, and a host of informal factors, such as rumors, shaped one's choice when starting a family. Even though the wishes and desires of individual young people did matter in the question of choosing their life partners, and even though love affairs did take place and indeed resulted in lasting marriages, up until World War II such affections were either exceptions to the general rule or treated as wishes and suggestions rather than binding decisions, and were only taken seriously in cases of extreme disobedience of parents and tradition. The opinion of parents was immensely, if not primarily, important, as was the role played by rumors and public discourse in one's village.[1]

Up until the 1940s, and to a lesser extent thereafter, the initiative to marry always belonged to a man and his family while women never proposed or suggested marriage, even in the least formal way. The groom's parents chose potential candidates for the position of a bride, and then the groom was asked for his opinion and any preferences among those candidates. The groom's parents made it clear that although they discussed potential brides with their son, and his wishes and preferences were taken into consideration, a young man was not allowed to have the final say in such life-long decision and commitment.[2] Similarly, the opinion of a bride was provisionally important, in the sense that an obvious dislike or outright hatred was taken into consideration. In such cases the bride's parents rarely forced marriage upon their daughter and justified their unwillingness to enforce their wishes by acknowledging that "not we, but she will spend the rest of her life with this man; it is her will to make a decision."[3] But the practice of acknowledging a bride's wishes led outsiders to misinterpret the freedom of choice that these women had. The custom of asking a bride for her opinion was more symbolic and ritualized than realistic and was rarely practiced,[4] not least because the traditional social norms of a Russian village required "an unquestioned respect for the elders" and only few young women ever dared to question the choice of their parents.[5] On all occasions, without any exceptions, the father of the bride had to approve his daughter's marriage and her choice of son-in-law. Village life demanded that young women complied with an image of a respectful, patient, rational, obedient, and pious young woman. Obedience to parents' choice was often the single most important factor in portraying this image, and hence in choosing a spouse. To keep "the face of the family," many young women practiced self-restrain and acted against their own wishes in favor of those of their parents, and as such these women argued that their choice—to respect their parents' wishes—was voluntary. Olga Kalinova from Damanovka farm of the Volgograd region, born in 1902, remembered:

> Once, around the start of the spring holidays (*maslenitsa*), my [future] father-in-law came in his padded jacket to arrange my marriage to his son Pavel. I worked for them doing some harvesting, weaving, sowing, and embroidery work. I am very skilled and can do everything. Since my mother was not at home, [my father-in-law] ordered my brother to call for me . . . I came and then they started to negotiate my marriage! Then the [future] father-in-law said: "See, I came to marry you off." Usually, [the future in-laws] send people to arrange a marriage but he came by himself, the father of Pavel. He told me: "See, Olga, we have wanted you [to marry our son] for a long while." But I resisted and argued that I had no dowry to take with me. I said: "I spend most of the day elsewhere and only use what is not mine. But when I get home I have nothing . . ." And my [future] father-in-law said: "You will have everything." But I resisted nonetheless, refusing to submit to this marriage. I said: "I will not marry." But they only kept quiet. At that time, everyone married like that; the bride says "no" but then her father comes in and says: "If you insist, then you are a bad daughter, hence leave the house, you are no longer our daughter." You see how things worked? You see how strong parents' will was? And then my marriage was arranged, and they brought Pavel, my groom, [into our house].[6]

Stories such as this are numerous, but they only provide an incomplete picture as other options, although not as common, existed in the village life. On many occasions, parents had no objections to their children's choice, and children had no objections to their parents' choice and saw it as sensible to follow advice from people who had the wisdom and experience of old age.[7] Other women were simply exultant to get married for the sake of marrying, no matter who the spouse was, especially if a woman's reputation was tainted by rumors or a previous marriage. Rumors of love affairs, premarital sex, and pregnancies or abortions could ruin not only the life of a young woman but also her entire family, and on such occasions most prospective grooms were treated as God's blessing by all family members, including the brides themselves. Returning to a parents' house after a divorce could also significantly taint a woman's reputation, and she was constantly reminded that she and her child(ren) were a burden on the family. Once again, in situations such as this, marriages were commonly arranged but widely welcomed by everyone involved. For these "second-quality wives," a marriage was not about love or affection, let alone common interests; life in general was a matter of mere survival, and a marriage was seen as a way to improve one's chances.

On the other hand, emotional attractions and a lack of parental involvement in rural marriages should not be overly romanticized either. Some women bitterly resented their choice and blamed their personal failures on their own inability to recognize the importance of parental advice. Typical of this group of women, M. Shilkina complained:

> I got married in 1935. I was born in 1916, and [my husband] Egor Ivanovich in 1914. I was not yet 19 year old, and [for various reasons] we could not register our marriage right away, so we lived a year together without a registration. Then we went to Tarbeevka, where there was a regional *soviet*. We came to marry, to register our marriage, but we [coincidently] had the same last name. They asked us there: "So, are you a brother and a sister?" We looked at each other: "What is that?" But then my father married me off to Egor Ivanovich [as I wanted]; after all [Egor] courted me for 4 years. My father was a team leader [at work] and knew Egor well, respected him. [Egor] was a good worker; [he was] the tallest, the strongest. I was also good, very plump. Egor Ivanovich worked in *kolkhoz* and was obedient . . . But then, after we married, I got very upset. His poverty was extreme. He could not even eat to fill his belly full! His father died early. His brother, Fedor, used to drink, and only his mother was left to care [for the family]. I came to their place and I was so sad. Their shack was tiny, the floor made of dirt. In my father's house, even though there were nine children, we lived better. Then we also started to quarrel with [my husband's] brother. I did not think hard enough when I married, though my father opposed at first . . . I did not think that [Egor] had not served in the army. Hence I was left alone as a soldier's wife . . . I still cannot understand why I did not think of all this [ahead of time] and still married Egor.[8]

In addition to parental wishes, the power of rumors was a major force to be contended with. The life of a village was enclosed and confined by geographical boundaries, but it was also public and even communal within the physical space of each village and among its members. Personal likes and dislikes, love affairs and fights were not simply a private matter but topics of discussion and debate among all members of the community, something to spread rumors about. As rural residents believed, each and every member of the community knew everything about everyone else, and privacy was an illusion that never found its way into real rural life.[9] In the times before television and even radio, the complete absence of any entertainment or social life (in the present-day sense of the word), in places where nights were dark without electricity and winters made even limited travel impossible, any— even the most minor—personal problem or affair was immediately sensationalized and made into an event of all-village caliber. The private lives of others attracted significant attention from everyone who craved a break from the daily routine of monotonous rural life. As a community, the village knew who was friends with whom, almost since the day they were born, and there was a common knowledge of mutual sympathies or hatred. Almost universally folk songs about marriages included rhymes attesting to the power of rumors:

> My dear, I am yours,
> Listen, my honey,
> Listen, my darling,
> To what people rumor about us.

Or, in another rhyme,

> [He] worked hard
> Gardening and fixing the fence
> Yet when [he] fell in love
> It was for everyone to rumor about.

The village community often lived according to social norms that emphasized strict supervision of any encounters between unmarried men and women. Youngsters were allowed go to get-togethers and dances when they were available, or, at a later time, Soviet clubs. Yet young women cherished their reputation of being "proper girls" and feared rumors of indecent behavior. It was unthinkable even for best friends of the opposite sex to enter each other's house without direct supervision of parents (or better not at all) for fears that such "indecent behavior" could generate "improper rumors."[10] One of the respondents in a recent sociological survey remembered that in the days of his youth, going back to the 1920s, there was nothing worse than premarital sex because "to go beyond a certain point was forbidden. [If rumors circulated], then the gates of this girl's house were marked by [human wastes] and no one married her." This respondent fully approved such ideas and thought that a lack of premarital engagement of any sort (even socializing) is the best option for a girl to safeguard her reputation.[11]

Once the rumors and parents sealed the destiny of a young rural woman, there was no going back. The anticipated outcome was a marriage, and many young women chose to concentrate on their dowry rather than discuss their prospective husbands. In the Russian context, the tradition of dowries goes back to the times of Peter the Great and the early eighteenth century. Peter I issued laws that prohibited women from inheriting any estate or possessions. Hence a dowry was the only way for parents to pass any material possessions on to their daughters, and for a groom's family to receive anything to supplement their own investment in a new family. The early Soviet legislation left no legal room for dowries, but the practice nonetheless continued to exist well into the Soviet era, and arguably in some locations to the present day. But the amount of the dowry, and the form and shape it took, largely depended on the circumstances of each family, as well as the socio-political realities of Soviet life. In the 1920s and 1930s, rural residents were so impoverished that they often ignored the tradition of dowries and only gave daughters "what [one] could get," "whatever was available," or even pots and pans the family had at hand. Very few families could afford any special items to go on a dowry list. By 1933, when famine raged across many rural locations in southern Russia and across Ukraine, women remembered that the only acceptable dowry was food, and the first to marry were girls whose families could boast a few kilograms of wheat flour as a dowry.[13] Yet as the rural life improved in subsequent decades and the villagers started to have more financial resources at hand, the system of dowries was revived and even informally codified. Depending of a family's position, the bride's parents were expected to buy (in the descending order of importance, based on their financial well-being) either a wardrobe, a bicycle, or a sewing machine. But at all times, brides were always expected to bring bed linen, curtains, table cloths, towels, napkins, pillows, and other household objects into their marriage. In various villages across the Soviet Union, a dowry was put on display to showcase a potential bride, and some families with daughters invested all their financial resources, to the last penny (or rather last *kopeika*), into such displays to make sure that "they were no worse than everyone else."[14] In select rural locations, the practice has survived to the present day.

Yet if rural marriages in the 1920s and 1930s were largely unaffected by the Soviet propaganda of gender equality and free love, the attitudes to marriage started to change after World War II. Starting in the late 1940s, most women cited romantic love as a necessary precondition for a marriage. In the war-torn countryside, where able-bodied men of a marriageable age were a great rarity, this quest for love was no less pronounced than in large cities that offered better prospects of finding a partner. At the very least, the quest for such love, if not love itself, was the first step towards sealing a marriage. One woman, Pelageia Aparina from Saratov region, recalled how she dreamed of marrying "a love of her life"; she dreamed that one day a young strong man, not a fairy-tale prince but a good village fellow in a clean and decorated village shirt, would come to her, extend his hand, and ask her to marry him. And according to her perception, things happened the way she dreamed about, although "a love of her life" was not in a silken shirt but a worn-out military uniform. They met at a local village party. A month later, she saw him walk towards her house, and her heart started pounding. When he knocked on the door of her house, she went out to him, not dressed as a princess in a fairy tale but in her apron, all the while wiping her soapy hands with her apron after doing laundry. He took one long look at her and asked her to marry him. The rest, as the old cliché has it, is history.[15]

But for many, this illusive quest for romantic love could not bear any fruit. Although demobilization in and after 1945 partially corrected the gender imbalance that existed across the entire Soviet Union during the war, the situation in the Soviet countryside remained depressing for rural women when it came to finding a potential suitor and life partner. In the post-war Soviet village, there was only one man in his twenties for five women of the same age group. Only in the late 1950s did the situation start to improve.[16] Yet many rural women, no matter how pressured they were to marry *any* eligible man rather than search for a love of their life, still chose to marry only for love and affection. Some suitors were rejected on the grounds of incompatibility, while even the hastiest marriages were concluded only if a bride "felt right" about it.[17] One respondent to village interviews remembered that she married in the early 1950s after knowing her husband only a few days. Yet, according to her own words, she agreed to marry him because everyone talked of him as a good man, kind and hardworking, and because he could potentially become the love of her life.[18]

The greater security and growing prosperity of the 1960s and 1970s (compared to the 1940s, for example) only deepened this quest for love, although some gender imbalance was still evident in the countryside.[19] The quest for romantic love was duly reflected in various surveys conducted in the Soviet countryside in the 1960s and 1970s.

Table 9.1 Primary motives in deciding to marry (% of all respondents)

Motives	Men	Women
Love	39.1	49.6
Similar interests, views of life	26.1	28.5
Sense of loneliness	14.1	4.7
Felt sorry/compassionate for prospective partner	7.4	3.1
Possibility of having a child	6.7	4.3
A mere coincidence	4.0	2.4
Financial well-being of a prospective partner	–	3.1
A potential partner had housing	2.0	1.2
Other motives	0.6	3.1

Source: S.I. Golod, *Stabil'nost' semii: sotsiologicheskie i demograficheskie aspekty* (Leningrad, 1984), p. 26

Yet, even with this overwhelming preference for love as a primary factor in deciding to marry, there was some room for practical concerns and considerations when it came to choosing a life partner. There is no doubt rural women believed that "one must treasure and protect one's family." Yet this statement implied different things for different people. Many younger people believed that only love and similarity of interests could cement a family, whereas their parents and older spouses assigned a greater role to factors such as mutual support, mutual participation in childcare, and a man's ability to provide financially for his family.[20] Recent research findings on successful and unsuccessful marriages in Russia showed (and these general findings were confirmed by many international experts in other settings and countries worldwide) that marriages for love were less stable and satisfactory to partners than marriages for material gain or those that were advised by parents. Thus, 43.5 percent of all marriage for love ended in divorce, whereas only 35.2 percent of marriages for material gain and 26.3 percent of arranged marriages did likewise.[21] Statistics aside, the practice of consulting with parents never fully died out in the Soviet countryside, and this was especially applicable to socialist republics of Central Asia (although Central Asian societies remained somewhat an exception to the overall dynamic of family life in the Soviet Union and hence they warrant an independent discussion).[22]

By the early twenty-first century, however, the practice of consulting with parents nearly disappeared even in the most remote Russian countryside, where only 1 percent of all marriages are arranged.[23] Many women argue that the older generation has little understanding of the "new world," which has changed dramatically. Others, such as Antonima Mikhailovna Taryshkina of Saratov region, say that:

> before, old men were wise, not like right now. One cannot find wise old men these days! I will never forget the words of my grandfather who used to say that when you pick a wife, you have to look at seven generations of her relatives to see if anyone in her family was lazy or stupid. To take a wife is not the same as to put on shoes; you cannot shake her off. Now everything is easy: [count to] one to get married, [count to] two to get a divorce. Back then, once married, you spend your life together.[24]

Yet even in the twenty-first century, the marriage dynamics in the Russian countryside nonetheless continue to be different from those in larger urban centers. Only 10 percent of all married couples in towns and cities lived near each other before marriage, and over 5 percent of all couples met at a summer resort and had a very short courtship.[25] By contrast, two-thirds of rural couples grew up near each other, and knew each other since childhood, or had known each other for a long time.[26] Similarly, both men and women marry younger in rural Russia than in urban centers, which sociologists often attribute to lower educational levels among rural youth.[27] In rural Russia, most men marry after they return from a mandatory military service (a 2-year commitment required of all males who turn 18), or on average between the ages of 21 and 25. Most women, however, marry when they are 18 or 19 years old.[28] By the time women reach the age of 27, they nearly universally have a husband and/or child(ren). Marriages with a significant age difference between spouses are almost non-existent in the countryside, and at most the spouses are 4 years apart. Socially mixed marriages where educational levels and incomes diverge greatly are also common in villages but rare in urban centers, especially when women are better professionally trained than men. In villages, it is common to come across families where the wife is a medical

Table 9.2 Criteria of happiness, female respondents only (in %, although respondents were allowed to have more than one entry)

	Highly important	Matters somewhat	Does not matter	Cannot answer
Children	67	11	1	21
Personal freedom	21	34	17	28
A job that I like	46	27	4	23
Comfortable living; good housing	62	15	2	21
Family	66	11	1	22
Love	58	18	3	21
A husband	57	18	4	21
Girlfriends	18	40	19	23
Career	5	19	50	26
A car	6	17	51	26
A countryside cabin	7	27	40	26
Domestic travel	8	33	31	28
Foreign travel	7	24	38	31
Favorite hobby	28	39	7	26
Fashionable clothing	15	44	15	26
Knowing important people	7	22	37	34
A combination of family and career	46	25	3	26
A combination of family, career, and a comfortable lifestyle	46	23	3	28
The man I love	50	18	5	27
Post-secondary education	21	37	15	27
To have all necessary things without working	8	11	31	52
High salary	35	36	6	23
Entertainment	7	17	32	44
Be helpful to others	36	36	4	24
To be liked by others	35	41	8	26
Conflict-free family life	41	27	7	25
Self-confidence	47	28	8	23
Success with men	12	28	28	32

Source: The survey was conducted in eight Soviet Republics and 16 regions of the Russian Socialist Republic (future Russian Federation) in 1990. *Zhenshchiny i demokratiia: Obshchestvennoe mnenie po aktual'nym sotialno-politicheskim voprosam (Rezul'taty sotsiologicheskogo isledovania)* (Moscow, 1991), pp. 85–6

professional (often a doctor) while the husband is a tractor driver, or the wife has college education and works as a teacher while her husband dropped out of school and is employed as an unskilled worker.[29] Marriages where men have a significantly higher social position, e.g. a wife milking cows and a husband working as an engineer, are not as common in the countryside.[30]

Sociological research demonstrates that when it comes to choosing a life partner, most young women look for men who are intelligent, can treat a woman as a friend, care about family and children, are hardworking, do not drink, and have a sense of humor. Physical appearance comes at the very bottom of the list. Men, on the other hand, look for women who are good looking, and also loyal, honest, intelligent, caring, kind, and good housekeepers.[31] But for both men and women alike, the formula of happiness seems to include children.[32] "Happiness" is the most elusive category of all to measure, but Table 9.2 summarizes some of the common aspects identified by women as a key (or keys) to happiness.

10 Conflicts and divorces

Love is beautiful but one cannot destroy a family for it.

The inconsistencies of the law, especially in terms of recognizing de facto marriages and allowing for an easy postcard divorce, make comparing divorce rates in pre- and post-World War II rural Soviet Union almost impossible. According to first-hand accounts, rural women in the 1920s and 1930s believed a divorce was a rare occasion in the countryside; they argued that rural women were patient, had few romantic notions, and did not seek a life that was different from what they, their mothers, their grandmothers, and all prior generations had. Moreover, a large number of de facto marriages in the 1920s and 1930s make tracing the "divorce" in such relationships extremely challenging. Yet the changes in the divorce proceedings and the refusal to recognize de facto or common-law marriages in the 1940s made official statistics more reliable for appreciating the stability and longevity of rural families. Accounts of women who are still alive and have a vivid memory of their marriage experiences in the 1940s and thereafter make the story of conflicts and divorces in the post-war Soviet Union more complete and, as a result, more accurate. Hence it seems justified to investigate divorces and their grounds starting in the 1940s.

The legislation was arguably the main force in shaping the official divorce dynamics, as many people chose to separate without filing for official divorce when the process became too costly or time-consuming for them. The ease of obtaining divorce prior to 1944, allowed for an unprecedented number of divorces to take place. But although the tendency for a high divorce rate changed after 1944, even legal restrictions and regulations that complicated the divorce process could not prevent a relative increase in subsequent decades in the number of couples who desired to part.

Both in the countryside and in urban cities, most people who chose to divorce were young. Most divorces took places among men age 30–39 and women aged 25–29. In the 1940s and 1950s, these age groups comprised two-thirds (66 percent) of all divorcees. Most marriages fell apart within 1 to 4 years of marriage (33 percent of all divorces) or after 5 to 9 years (another

Table 10.1 Divorce rates in the USSR, 1940–55, in thousands

Years	Overall	City	Village
1940	205	108	97
1943[a]	83	38	45
1944[a] (January–June)	*58*	*28*	*30*
1944[a] (July–December)	*11*	*5*	*6*
1944[a] (Total)	69	33	36
1945	6.5	5.9	0.6
1946	17.4	15.9	1.7
1947	28.4	25.4	3.3
1948	41	35	6.0
1949	55.9	46.7	9.2
1950	67.4	54.8	12.6
1951	76.7	62.0	14.7
1952	86.1	68.7	17.4
1953	98.4	79.5	18.9
1954 (January–June)	*55.9*	*47.0*	*8.9*
1954 (Total)	114.4	97.0	17.4
1955 ((January–June)	*63.6*	*57.2*	*6.4*

[a] without occupied territories

Source: *Sovetskaia zhizn. 1945–1953* (Moscow, 2003), p. 697

33 percent of all divorces).[1] Rural families were in the minority when it came to dissolving marriages. Most people in the countryside condemned divorced people and blamed the divorcees for their personal failures and for their "unwillingness to work hard enough" at reconciling any differences or solving problems.[2] The most common reasons for divorce in the Soviet countryside were the abuse of alcohol, commonly by a husband; significant interest and lifestyle differences between spouses; and crime. In the 1940s and 1950s, it was mostly men who initiated divorce in rural families, although the tendency changed in later years when women took the lead in the 1960s and there-after. Many rural women embraced the notion that men were breadwinners and protectors of families and hence women had to tolerate anything and everything for the sake of keeping *a* man in the house.[3] The overall sense among rural residents that divorce was not a solution, and should be condemned, persisted to the end of the twentieth century; even in the 1980s urban families dissolved their marriages twice more often than rural couples, although, to be accurate, we need to remember that the rate of divorces in the countryside went consistently up, even if it failed to catch up with the urban dynamics.[4]

In the 1980s, 2.6 million marriages were registered in the Soviet Union, paralleled by over 1 million divorces. The divorce curve was continuously on the rise for most parts in the second half of the twentieth century; for example, the number of divorces grew from 757,000 in the 1970s to nearly a million in the 1980s (for Russian Federation only, see Table 10.2). As a result, over 12 percent of families were single-parent families, while 829 of 1,000 couples who divorced had produced children. In real numbers, over 13 million people were divorced in 1989 (of them 8 million were women), and as a result of new divorces legalized in 1989, 775,000 children joined the ranks of those living with a single parent, almost universally their mother.[5] The statistics became even more depressing in the 1990s, when the number of marriages declined from 1.3 million in 1990 to 1 million by 2000, whereas the number of divorces went up from 500,000 to nearly 763,000. Divorces in rural regions increased 1.8 times, while the urban locations only saw an increase of 1.3 times, although they had a higher starting point.[6]

Although some divorcees remarried, sociologists argue that the relatively low number of second and subsequent marriages indicate that many divorcees preferred to live in unregistered relationships, commonly without the fear of poor treatment of children by a step-parent.[7] Concerns for children's welfare, however, were not a significant factor in choosing to remain married or divorcing.[8] Only a small number of families postponed their decision to separate because they had young children and did not want to force their children to undergo the massive stress of seeing their parents' divorce, and

Table 10.2 Marriages and divorces

Year	Marriages	Divorces	Marriages	Divorces
	(thousands)		*(per thousand people)*	
1960	1,500	184	12.5	1.5
1970	1,319	397	10.1	3
1980	1,465	581	10.6	4.2
1990	1,320	560	8.9	3.8
2000	897	628	6.2	4.3
2001	1,002	764	6.9	5.3

Source: *Zhenshchiny i muzhchiny Rossii. Statisticheskii sbornik* (Moscow, 2002), p. 38

Table 10.3 Second and subsequent marriages, 1980–95 (% of all marriages in Russia)

Year	Men		Women	
	Remarriages	Of those, after divorce	Remarriages	Of those, after divorce
1980	18.9	16.1	17.9	14.3
1985	23.8	20.7	24.6	20.1
1990	25.3	22.3	24.6	20.4
1995	27.5	25.0	27.4	23.3

Source: *Naselenie Rossii 1996. Ezhegodnyi statisticheskii doklad* (Moscow, 1997), p. 68

to be left with only one parent. Ironically, when women saved their marriages "for the sake of children," many of those children came to resent their parents for failing to divorce; these children argued that seeing parents constantly fight or having a parent who was an alcoholic was even more stressful than a divorce would have been.[9]

Although the proportion of remarriages to all marriages was not unique to the Soviet Union and later Russia, as Table 10.4 of comparative data for some select countries indicates, it was highly unusual in relation to the number of divorces per thousand people.

The grounds for divorce were, and continue to be, numerous and complex. In the first few years after marriage, most relationships fell apart as a result of personal incompatibility and financial hardships. Yet divorces after 5 to 9 years of marriage, which made up one-third of all divorces, had more diverse reasons. They included grounds such as "moral decline" (most often meaning alcoholism); improper behavior of a spouse (i.e. adultery); cruelty and physical abuse; neglect of spousal and parental responsibilities; short temper; and

Table 10.4 Divorce and remarriage data for selected countries

Country	Divorces, 1985 (per thousand people)	Remarriages, 1980 (% of all registered marriages)
Japan	1.38	16.9
Belgium	1.86	15.0
France	1.95	11.4
Italy	0.29	16.6
Switzerland	2.37	20.6
Great Britain	3.20	23.7
USA	4.96	32.5

Source: *Chelovek i trud*, 1992, no. 3, p. 8; *Narodnoe khoziaistvo SSSR v 1990 g. Statisticheskii ezhegodnik* (Moscow, 1991), p. 83

Table 10.5 Grounds for divorce

	%
Illness of a spouse	2.8
Infertility	2.5
Forced long-distance separation	7.7
Conflict with step-children	0.5
Physical abuse	3.8
Infidelity	17.5
A spouse has another partner/family	21
Alcoholism	14.2
Personal incompatibility	13
Parental involvement and conflict	3
Love for another person	7.5
Religious and social incompatibility	6.5

Source: N. Soloviev, "Razvod, ego factory, prichiny, povody" in *Problemy byta, braka i sem'i* (Vilnius, 1970), pp. 123–4

personal incompatibility. Also common were infertility, jealousy, repetitive fights, conflicts with relatives, and unfair division of household duties and chores. Even personal character traits—egoism, stubbornness, suspiciousness, toughness, and inability to love—made their way into this long list.[10]

In the late 1960s and into the 1970s, the initiative to end the marriage passed onto women. In the countryside, although the village community still assigned a degree of blame on divorcees, some women started to voice their concerns that not all marriages were worth preserving.[11] This concern was shared by many women whose husbands became alcoholics. As a matter of fact, in over 40 percent of all rural divorces the main reason for divorce was the husband's alcoholism. In many other cases, the conflicts and physical abuse that were cited as grounds for divorce, were also related to drunkenness.[12] L. Bukareva from Tulsk region wrote to a women's journal, *Rabotnitsa* (*A woman-worker*), in 1965:

> I have four children. There isn't a day when we do not have a fight in the house. My husband drinks away all wages, all bonuses. I went to the place where my husband works but to no result. There are many families like ours in our village.[13]

Similarly, Olga K. sent a letter to another woman's journal, *Selskaia molodezh* (*Rural Youth*):

> It became difficult for me to live and make sense of my life. The problem is that my husband drinks. He loves me when he is sober but once he starts drinking, then he forgets everything, he does not come back home for days at a time until he spends all the money on drinks. I know that there are many families like this, and it is difficult to meet a happy family this day and age.
>
> I also grew up in a family where my father drank, and I do not want my children to have such a life. Yes, I felt sorry for [my husband], I thought he would suffer [without me], but now, as time passed by, my pity turned into anger when he is drunk. I do not know if I have guts to file for divorce but I cannot live like this anymore.[14]

Drunkenness was not a new phenomenon to rural Russia. But in pre-revolutionary Russia, only religious holidays, significant personal events such as weddings and funerals, and the end of the harvest were the reasons to get drunk. Ironically, all these traditional celebrations survived throughout the Soviet era, but new nationwide Soviet holidays were added to the list of occasions to drink for, as well as many *kolkhoz*-related events. Many local party officials complained that "in various regions and collective farms . . . religious holidays [were] accompanied by collective drinking sprees and subsequent fist fights and accidents."[15] But whereas most state officials cared little for each worker and were only outraged when drinking happened during

religious celebrations, many women saw their husbands first develop addiction
to alcohol and then start neglecting their family duties.

The Central Committee of the Communist Party issued a vague "Resolution
on Alcoholism" in 1957 aimed to combat alcoholism, justifying its actions
by the fact that "alcoholism had become endemic, and especially during
religious holidays."[16] Immediately, 777 collective farmworkers who were
"established to abuse alcohol" were prosecuted and 1,148 more of those
"whose actions were tainted by the use of alcohol" were charged with anti-
Soviet behavior. Most of these people were young men under the age of 27,
or young husbands and fathers.[17] But of course, such measures were more for
show than to create real change; they affected only a few people at the time,
while male alcoholism was indeed becoming endemic nationwide. For factory
managers and local officials who had to deal with alcoholics on a daily basis,
the most common way to sober these men up was to fire them from their jobs.
Yet for most addicts, such measures had the opposite effect; emotionally
devastated and lacking even the most rudimentary discipline their jobs used
to provide, "functional alcoholics," or those who used to sober up at least for
some workdays, started to drink more rather than less. One rural woman's
story exemplified the situation:

> Where do I turn for help? My husband, a communist, drinks to no end.
> I am bruised all over; he attacks children with a knife. My younger son
> has become nervously ill [acquired stress-related psychological disorders].
> I went to the party cell where my husband belonged but they laughed at
> me. I went to the village communist committee and there they phoned
> his job . . . He got fired. After that he got completely wasted.[18]

Many women failed to secure any help from state institutions in safeguarding
themselves from the physical abuses of their drunken husbands, and some,
fearing for their lives, were even forced to flee their villages. Nina M., for
example, had to leave her work in a small Siberian village and move to a
collective farm in Central Asia with her elderly mother and a small child. She
wrote to one of the women's journals:

> I have a tragedy in my family. My husband drank, debauched. And I was
> forced under the circumstances to hide away at our neighbors' places with
> my 70-year-old mother and a child. At work, I was respected. But at home,
> I was a helpless victim of [my husband's] tyranny. My neighbors and I
> went to the police, to the party cell, to court. No one wanted to help me.
> In the police, they told me: "*You* got into this, *you* get out of it. Come
> back when he kills you."
>
>   How long can we tolerate this drunkenness and this abuse of women?
> Where is the solution? No one is there to help. Isn't it time we find a
> solution instead of making women with children flee their homes? . . .
> To tell me that it is my "private" family business is to avoid responsibility.
> This is by far not my private and not my family business.

If the Soviet state does not protect us, women-workers, who will? Women who live in a constant fear of death from the hands of their husbands (and I am not exaggerating) do not write to you; they do not write anyone as they are convinced that they lack any protection. And men-alcoholics, knowing that nothing will be done to them, kill all that is cheerful and optimistic in the hearts and souls of their wives, [who are also] mothers. But it is we, women, who work no less than men, and we also raise children and study [to improve our skills].[19]

Alcoholism, of course, was not the only cause of failed marriages. The second most common reason cited by rural residents who sought the end of their marriage was incompatibility. This term, however, covered a wide range of family conflicts and problems, including those related to the abuse of alcohol. The analysis of court cases demonstrates that this term often included the complaint that a family lacked children. However, it was a refusal to have children rather than infertility that led to divorces in such cases. Most commonly, women explained that they were fearful to have children with husbands who drank habitually, or that they refused to have children with a man who repeatedly cheated on his wife. No less common was the refusal to have children when the woman lived in the same housing as her in-laws and had a problematic relationship with her new relatives.[20]

Some spouses also cited the unfortunate financial situation of the family as grounds for divorce. Many women explained that before getting married, they believed that a man had to be a breadwinner, and that a marriage had to result in an improved financial situation for them personally. The reality, however, had proven otherwise. As a result, even though most women worked and most men could not earn wages beyond those set by the state, disappointments with men's income abounded. Young women such as Lidia and Zinaida shared into this disappointment:

We think that the old proverb *all you need is love* is outdated. One cannot be fed and clothed by love. And love, even if you have it, at best lasts only until the wedding day. The main aspect of family life in our rational twentieth century is material. Tell us: How can a woman marry the man she loves if he only makes a hundred rubles a month? [average young worker's wage] How will he provide for the family? ... We are just normal, average, modern girls who are reasonably skeptical. In brief, [we are] quite typical, and we express the opinion of the masses.[21]

As women took the lead in initiating divorces, they also became less willing to search for compromises. Thus, according to various surveys, when divorce was looming on the horizon, 44 percent of men were willing to do anything to avoid conflict; at the same time, 33 percent of all women in such a situation avidly opposed any attempts to remedy the situation, while the majority of the rest of the women admitted that they did not care to search for a compromise.[22] Scholars believe that, at least partially, the physical and emotional burdens women had to carry, especially in the countryside, were responsible for this attitude among women.[23] Work in the collective farms continued to be for most parts manual (at best only 30–40 percent of all labor was mechanized), and it was men who had access to tractors and other equipment while women continued to perform manual tasks.[24] A. I. Red'kina from the Vologda region, recalled:

> Right after I stopped going to school I went to work for a cattle yard where I spent nearly a year. [After leaving this job], my job was to go around the local villages and sell milk that I received [for sale] from two nearby cattle yards. Besides dragging the containers of milk to my customers, I also was responsible for washing the containers, [something that was done] manually. My wage was very low, even though I had to distribute over a ton of milk daily [approx. 300 gallons] . . . Then I [switched jobs and] went to milk cows. I had 25 cows and the work was difficult; we milked cows by hand, we washed them, guarded them [from being stolen], shoveled the manure, took cows to the fields to pasture, [and] in the summer we even took cows to pasture at night; [we also] had to scramble for additional hay to supplement our cows' [officially-distributed] ration, and when things were especially bad in the spring, we searched the fields for any leftovers from previous harvests.[25]

Demands of agricultural jobs as well as work in personal garden plots, household chores, and childcare responsibilities left little energy to these women to care for reconciliation, and little time to invest into getting medical help for their husbands with substance-abuse problems.

Yet if alcoholism was a common reason for dissolving marriages, in rural areas love affairs were treated as insufficient grounds for divorce. The village community recognized that people change over their lifetime and affection, even when it existed to begin with, dies out, and hence spouses can find themselves attracted to other people. Yet a family was more important than "fussing over" your spouse's affair. Even in the late 1980s, over 70 percent of married people in the countryside (and 60 percent of those not yet married) believed that the family's indivisibility was the most important thing, indestructible even by cheating.[26] We need to keep in mind, of course, that in enclosed, small rural communities, affairs could not be kept secret, and hence many deceived spouses were fully aware of their position. Many women also confided in their girlfriends that they loved someone other than their husbands but were not willing to divorce so as not do their family any harm. "Life is very challenging," summed up one woman:

> I met a man when I was younger; now I am 40 years old and he is 2 years my senior. We dated for 3 years . . . Then the time went by. Now he has a family and two sons. I married likewise. My husband is good; we spent 20 years together . . . but I dream of that old boyfriend of mine all the

time. I have no feelings whatsoever for my husband, even pity. It is all about duty, a duty of a wife to her husband. Everything I do, everywhere I go, I only think of the [man I love] . . . I want to do everything for him and him alone . . . But my life has treated me harshly twice, by making me live with a man whom I do not love but to whom I owe everything for his care and his kindness; and by making me love another man.[27]

Responsibility for children, regardless of a woman's suffering and her love for another man, was cited as of primary importance over anything else.[28] Although women shared into this notion of putting family above love more commonly than men,[29] some men also put duty and responsibility above personal affections.[30] Yet city life did not welcome such self-sacrifice. Women who were intimately familiar with the rural norms of behavior (either because they had relatives in the countryside or because they were born there) often looked at such marriages with astonishment and saw it as an archaic appendix of a simple rural life.[31]

Rumors, and fears of them, also shaped family dynamics and popular perceptions of divorce in villages. Women who had affairs with married men were always condemned by the village and were openly criticized, whereas the popular support and sympathy always belonged to the women who were betrayed (even in cases when separation was mutually agreed). The public opinion universally shared an image of a proud and undeniably "better" wife who was left by her husband only because of the "evil" deeds of another woman. Those evil deeds ranged from overtly displaying affection for a man, and thus seducing him, to performing magic spells to charm a man away from his wife. After the rumors of a love affair spread throughout the village, the entire community typically went on a quest "to defend" the family, a struggle in which all means were justified. Women in these relationships had to listen to "trustworthy" advice from relatives, friends, and neighbors, or at times even came under public scrutiny during collective party cell meetings at work.

Because village life was so tightly knit, everyone was in the habit of gossiping about everyone else, and many women who found new passion had no one to turn to for support. Most commonly, their girlfriends and confidants betrayed their trust and started to rumor about the intimate details of their friends' private lives. Rumors were taken so seriously that even unjustified rumors often broke families and spoiled lives.[32] On the other hand, wives had many resources at hand to "discipline" their husbands (although their efficiency is questionable), especially since communist party cells were on their side. A wife could complain about her husband's love affair to a local village party cell, to his work's party cell, or to a great number of district, regional, and even federal party organizations. The response followed immediately. Men who were called in for a "private conversation" with the party committee members were often tainted professionally for life, or even lost their jobs as they were labeled as "socially unreliable elements." In the more extreme cases, a man could get a reprimand written into his party membership card or even lose his party membership, something that caused definite life-long damage to one's reputation and professional career. Proposals "to save a lost communist" and "to separate my husband and his lover by forcing her to relocate" reached even the Central Committee of the Communist Party of the Soviet Union![33] Precisely because of the severe consequences such measures entailed for estranged men, they were often successful in officially reuniting families. Of course, whether these measures could repair emotional damage done to all parties involved is questionable.

Table 10.6 Who women turn to for emotional support when they experience difficulties (of all sorts) in life (in %; women only)

	Most often	Sometimes	Never
Girlfriend	20	39	12
Husband	40	16	7
Friend	9	33	17
Co-workers	10	43	15
Church	2	9	43
Neighbors	2	22	37
Fortune teller	1	8	49
Father	11	20	23
Mother	36	30	11
Children	22	32	15
Professional unions	4	27	44
Party committees	3	16	55
Local party cell	2	12	57
Women's party organization (*zhenotdel*)	3	13	53
Rely on myself	56	11	2

Source: *Zhenshchiny i demokraiita: obshchestvennoe mnenie zhenshchin po aktual'nym sotsial'no-politicheskim voprosam* (Moscow, 1991), pp. 84–5

11 Domostroi

An obedient wife is as pleasing as a beautiful necklace.[1]

Domostroi, or literally translated as 'domestic order', was a book compiled during the times of Ivan the Terrible in the sixteenth century to outline every detail of household management and family relations. Although very few people in Russia used the word domostroi to refer specifically to that work, domostroi remained a key phrase to depict the patriarchal relations that prevailed in most families up to the twentieth century and beyond. The system of absolute and complete female subjugation to her husband—socially, religiously, and economically—was often reinforced both by law and husbands' fists, whose beating most women accepted as a normal and routine part of their lives. The original *Domostroi* encouraged wife beating, with some limitations:

> If a wife or a daughter does not obey orders, [if women] do not fear [men], do not do what their husbands or fathers tell them, then whip them with a lash according to their guilt, though do so privately, not in the public eye. [One cannot hit] in the ear, or face, or under the heart with fists, or hit with feet. [A man can] whip with a lash in a way that is reasonable, and painful, and terrifying, and also so that people do not see it and do not hear it.[2]

As a matter of fact, even by the turn of the twentieth century, wife beating was still a routine practice. The customary law, unlike the legal code, did not recognize wife abuse as a crime, and most women were either not aware of their legal rights or were discouraged from seeking justice in courts. In pre-revolutionary Russia, most people on both sides of the gender divide understood "physical punishment" as an appropriate measure for reinforcing authority. Often men resorted to violence when they were under the influence of alcohol; jealously, rumors, even minor disagreements over running the household could trigger the beating. But at times men did not even attempt to provide an explanation for their actions; they took it for granted that "[women] need to be beaten; life becomes miserable if women are not beaten."[3] Men resorted to using anything at hand during such violent rages, including whips, sticks, boots, and shovels. Some men beat their wives to death and were subsequently persecuted and exiled to Siberia, but most men nonetheless were more restrained and perceived wife beating as an effective way to negotiate family frictions and reinforce their authority. Because of the commonplace recognition of the effectiveness of this measure, wife abuse continued and persisted in the countryside.

Occasionally, rural women had the knowledge and courage to take their cases to the *volost'* (provincial) courts, where male judges did not hesitate to apply law to family affairs, despite popular perceptions, and eagerly sought to shield women from domestic abuse. Men were often found guilty in these cases.[4] When the physical abuse did not result in any permanent physical damage to a woman's body, the most readily applied measure of prosecution was whipping (20 lashes) in order "to prevent evil deeds in the family and live in harmony."[5]

Yet for most rural women, going to court was not an option. Instead, some women relied on tacit protests such as running away to escape the domestic tyranny of a husband. By the late nineteenth century, women increasingly resorted to this measure. But running away was complicated by the fact that all Russian citizens needed passports to move around—few if any rural residents had passports, and women had to appeal to their local district court to get one. According to law, husbands had to agree to their wives getting a passport of their own, a permission that only very few men ever granted. Thus the women were forced to go back home, and if the runaway continued to hide out in her parents' house, she was returned home by force. Even regional courts of appeals routinely ordered women to go back home, although they also warned men again further violence to avoid persecution.[6]

The customary practice of wife beating, or at least the tolerant attitude to it, persevered even after 1917 when gender equality was pronounced. Rural men still considered that "if you don't teach a woman by beating, then you see no results" and resorted to the power of fists to express their jealously, to solve domestic disputes, and at times simply because a wife happed to be in the wrong place at the wrong time (most often implying drunken rages). Opinion polls conducted by anthropologists and social scientists in 1926 in one of the central regions of the Soviet Union demonstrated that nearly 60 percent of rural men considered wife beating the most effective way to solve disputes and reaffirm male authority over their spouses.[7] A similar poll conducted in 1937 revealed a curious blend of Soviet realities and customary practices. In this poll, men attested that they beat their wives when the two argued about joining the collective farm; when a husband and wife disagreed about listing their religious affiliation in the census questionnaire; and when the two had to decide about talking to local party authorities.[8] At the same time, rural women, at least those who came of age before the October Revolution of 1917, universally confirmed that both the fear of being beaten and the threat of domestic violence were an integral part of family relations. This attitude was so internalized that even in cinema, rural heroines personified this common aspect of a rural life. Thus, a well-known and widely publicized film of the 1930s, "A Member of the Government," set out to tell the country, in a typical propaganda mode, of the unique and enormous role that peasant women played in building socialism in the USSR. Its main heroine, a People's Deputy to the Supreme Soviet (played by Vera Maretskaia), when addressing the Supreme Soviet in Kremlin, started her speech with the following words: "and here I stand in front of you, an average Russian woman [*baba*], more than once beaten by her husband."[9]

But both men and women who belonged to a generation that matured at the time of the Soviet rhetoric of gender equality during the first two decades of Soviet power, started to progressively criticize such practices as contrary to the revolutionary and socialist ideals of the state. As well as this change, both law and prosecutors increasingly took the side of the women. Nonetheless, it proved nearly impossible to fully weed out the practice of wife beating that had been shaped and reinforced by centuries of customs and generations of shared mentality.

Nonetheless, attitudes started to shift, little by little. Many cases of violence against women such as wife beating, by some estimates as much as 60–70 percent, were done under the influence of alcohol.[10] But by the 1960s and in later decades, when it came to violent rages, some rural residents started to take it as their responsibility to act on behalf of victims. Precise numbers are not available, but anecdotal evidence suggests that rural residents were more likely than urban residents to speak up in defense of their neighbors and friends, and to mingle in private family matters. A typical example is presented in the following letter written in August of 1965:

> Our collective farm chairman got drunk on July 23, 1965, something that happens to him quite often, and then he beat up his wife, while swearing so intensely and loudly that he woke up the entire village. Then he got a hold of his gun and wanted to shoot his wife but fortunately, she managed to get out to the street where she had to stay till morning. [He] has been living an immoral life, especially when it comes to spending time behind the closed doors with one of collective farm women; [he] drinks vodka with her as well, then he goes to beat his wife over again, but [she does not do much because] after all she has four children.

To emphasize his amorality the letter added that the chairman even once took a bale of hay home without paying for it. Further investigation found the charges of taking hay false, but the prosecutors nonetheless announced that the chairman's actions toward his wife were "wrong" and he was "severely reprimanded."[11]

This and similar cases can also be read in their social context to demonstrate that the letter's author was dissatisfied with the chairman for work-related reasons. The cases of domestic violence were easier to prosecute if a man was also known for violence in the workplace or towards non-family members. But wife beating, as well as other forms of assaults on women, were left unrecognized in many instances, not the least because the Soviet legislation recognized as crimes only those acts of violence that led to severe physical damage. Other cases, however, were unrecognized by law, including marital rape, verbal abuse, bullying, and so on. The discussion of legislative changes to address such forms of violence did not come about until the 1990s, and even then the results were mixed at best.

Domestic violence reached a new high in the 1990s. In 1994, Chief Prosecutor's Office registered 565,300 female victims of violence, including 40,000 who suffered because of their husband's jealousy or minor disputes over domestic chores.[12] In the 1990s, nearly 12,000 women died at the hands of their husbands annually, and 54,000 more were registered as having suffered serious physical damage inflicted by their husbands.

Domestic violence was, of course, by no means limited to Russia. A total of 34 studies conducted in various countries demonstrated that 25–50 percent of all women (and according to some estimates the number is even higher) have been a victim of harsh treatment and abuse at the hands of their partner (current or former).[13] Especially troublesome is the fact that women who experience violence from domestic partners are more likely to develop psychopathologies and mental illnesses than those who do not.[14] Among the female victims of domestic abuse, the most common problems are depression, post-traumatic stress disorder, insomnia, and substance abuse. Women are the most represented group of victims to undergo psychiatric treatment and are more prone to suicide than others.[15]

There is a popular notion in Russia that a man beats his wife only under the influence of alcohol, or when he is pathologically jealous of his wife and the attention she receives from other men. There is also a stigma attached to "exposing your dirty laundry," i.e. talking about assaults openly or taking male partners to court. Often women believe that only families of low educational and socio-economic status are willing to "take the trash out of the house," and most women whose husbands occupy positions of prestige in their workplace consider it inappropriate to ruin a man's career over something as minor as wife beating. As a result, women tend to keep quiet, although studies that emphasize anonymity, such as the one conducted by the Center for Crisis Management "Anna," exposed that over 75 percent of all men hit their female partners at least once in their time together.[16]

Director of "Anna" shares some of the concerns and challenges she faces in running the center. On one hand, in recent years 70 percent of those who died as a result of domestic violence were women. The remaining 30 percent were male deaths, but nearly all these males died as a result of wounds inflicted by their female partners who attempted to defend themselves. But on the other hand, as Director Potapova argues, many women who turn to the Crisis Management Center for support face significant opposition from their friends and social pressure to withdraw all charges against their husbands and partners. They often hear:

> Do not do it! Think of leaving your children [without a father] and also that you will ruin your [children's] biography. How will they feel when in every questionnaire they fill out in their lifetime, they will have to write that their father is in prison?

Moreover, even when women decide to prosecute their partners, as a general rule, no restraining orders are issued, and both the victim and the perpetrator of violence continue to live in the same house, thus increasing the chances of further domestic violence and even fatal outcomes.[17]

In 2000, over 250,000 female victims of domestic abuse withdrew their charges against their partners on the grounds of "mutual appeasement." Ironically, even though most women at the time of the incident say that they see an immediate need to divorce their violent husbands, on average it takes seven attempts to leave the perpetrator of violence before the women finally separate from them for good.[18] Several other stereotypes work to the women's disadvantage. In Russia most women consider it better to be married at any cost rather than be alone (which is especially true for women with younger children who do not want to become single mothers), and sadly, in defense of their marriages, some women even aim to legitimize the beating and find excuses for their husbands. Cases when women refuse to call the police even when urgent medical care is needed are also common. A medical professional complained that all too often, a wife "beaten half to death" called the ER mobile units (i.e. 911 services) but refused to call the police and even refused to go to a hospital despite being urged to do so by the doctor! After "shedding waterfalls of tears," these women explained that "who else is going to take care of the house and do all house chores [if a woman goes to the hospital]?"[19]

For some women, the physical violence had really become ingrained into their relationship. Some women who condemn domestic violence in public later reveal in private conversations that they take it as an unfortunate but inevitable reality of life that men need to reassert their authority over their wives, and that wives have no choice but to confirm their submissiveness by tolerating the beating. Breaking through such stereotypes is ever more problematic if this is the type of relationship they witnessed between their parents when growing up. A writer and a feminist, Maria Arbatova, wrote that:

> Sometimes women come to me [to complain], 'my husband beats me! I am getting a divorce!' But I fully understand that they would never divorce. This is just their model of relationship. A woman like that would miss something, would feel that she lacks sometimes with a man who is normal. If her father always physically abused her mother, she would subconsciously look for signs of physical violence everywhere.[20]

Rural women fear and experience domestic violence more often than their female counterparts in large urban centers. Various factors explain such differentiation. First, most women in their late teens and early twenties are less willing to tolerate domestic violence and are more mobile than their older female relatives. But the demographic makeup of the countryside tends to lean toward women who are past these early years. The enclosed community of a village, where everyone knows everyone, also makes many women hesitant to share their concerns and problems. Indicatively, rural women who are divorced are twice as likely to admit domestic abuse from their former spouses as women who are currently married to violence perpetrators. Moreover, whereas about two-thirds of women nationwide think that violence is "hard to avoid," the number goes up to over 80 percent if cities are not included. Finally, rural families have a higher percentage of socio-economic problems and incomes below the poverty line, and thus violence that starts with disputes over financial matters is more common in (although by no means limited to) rural locations.[21]

Once again, many of these problems are not unique to Russia. Worldwide, only 45 countries of 193 have laws against domestic violence. The shortages and limitations of such legislative omissions are widely discussed. For example, they were vividly articulated by a famous lawyer, feminist, activist and essayist Gisèle Halimi, who, speaking to the Moscow public in June 2003, outlined the need to struggle for the right to abortion, greater political rights, and a need to combat domestic violence.

Violence aimed at women took other forms as well, most commonly in the form of rape and murder. Both were severely punishable by law but both, with significant infusion of alcohol, continue to take place. Ever since the times of Peter the Great in the early eighteenth century rapists, if identified, risked losing their heads or at best going into Siberian exile for life.[22] According to customary law in the countryside, rape was considered the worst of all crimes. A raped woman was an object of pity and was considered "marriageable," unlike young women who practice premarital intercourse. A rapist, however, was commonly ostracized, and even well-off men faced significant problems in finding suitable wives if they were known to have violated a woman. An underage rape was never overlooked or written off; such men were labeled as "equal to Satan" and had to reach a settlement with the girl's family, usually in the form of payment. Even then, some families sought justice in courts. For example, a governor's court of Tambov *guberniia*[23] heard 30 such cases in 1900, a number that was typical for prior years as well.[24]

Beginning in 1922, Soviet law progressively increased the severity of punishment for rape. In 1962, especially gruesome cases of rape became punishable by death.[25] It is worth pointing out that in the 1960s, nearly 70 percent of all cases of rape felt into this category, mostly because raping underaged women was considered a case of aggravated rape.[26] But the law, although good in its intentions, proved to be controversial. Many rapists, knowing that their crime was punishable by death, sought to conceal the crime, hide the evidence, and escape prosecution by murdering their victim. They believed that only death guaranteed silence, and such an attitude seemed to increase the likelihood of murder after rape. Judges were also reluctant to issue death sentences, and considering that most lawyers advocated that rape was "aggravated" only if it resulted in a death of the victim, only a small number of court hearings resulted in death penalty in the 1960s.[27] In 1996, the Criminal Code of the Russian Federation provided for a maximum of 15 years of imprisonment for rape, even in the gravest cases.[28]

But stereotypes and limitations similar to those discussed for wife beating could be applied to the issue of rape. Thus, fearing stigmatization and battling with various stereotypes, only an insignificant number of women (in proportion to rape cases) ever turn to law enforcement agencies for support and to punish the rapist. According to official statistics, rape and attempted-rape cases were only slightly on the increase and varied rather moderately throughout much of the Soviet history. Thus, there were approximately 3,700 cases in 1940; 3,100 in 1949; 3,400 in 1950; 4,000 in 1951 and so on, to approximately 3,000 cases annually in the 1980s.[29] But the statistical curve changed dramatically in 1990, when over 15,000 cases were registered, although the numbers declined to 10,900 in 1996 and 9,000 in 1998.[30]

Official numbers poorly reflect the reality of sexual abuse towards women. Most women never admit to being victims. The estimates for the actual numbers of rapes are four to five times higher than official data,[31] and criminalist V.V. Luneev estimates that in Russia the proportion of registered to real rapes is one to six. Comparatively, in the US the actual rape is 3.5 times that of recorded.[32] In some cases, victims of rape were forced to marry their violators when such women were no longer deemed marriageable, although such cases were rare. However, women were left to deal with the psychological trauma on their own. Even in cities the resources were, and continue to be, limited, and in the countryside there were no resources or support groups of any kind. Not every woman managed to handle the trauma, and some ended up becoming alcoholics or committing suicide.[33]

According to the Moscow-based center for victims of sexual abuse, only 2 percent of women who are subjected to sexual abuse ever turn to police for assistance.[34] Because on most occasions rape is committed by an acquaintance or a man the victim knows, the success rate of finding a perpetrator is high, estimated at roughly 90 percent. Yet the fear of being stigmatized and the continuous denial of such notions as date rape, marital rape, and sexual harassment, leave most men at ease to commit further acts of sexual violence against women.

On a final note, by 2009 gender research centers, small in number but crucial in their work, had already started important work on educating women about what constitutes sexual abuse and on collecting data (mostly by the means of survey) to expose the seriousness of the problem. For example, the Moscow Center for Gender Research demonstrated that in the first half of the 1990s,

25 percent of all women experienced sexual harassment in the form of demands for sexual services and intercourse in the workplace.[35] Once these women faced advances from their bosses, their two options were either to quit the job and risk unemployment or to satisfy his demands. At times, women intentionally used their physical attractiveness to provoke such demands for the sake of financial and career advancement. Yet most women in the 1990s, and progressively so into the present, became aware of notions of gender equality and a discrimination-free work environment. They want and demand a right to a work environment, career advancement, salaries, and collegiality that exist in a sex-neutral and professional space, separate from their physical appearances and their bodies.[36]

12 Alcoholism in the countryside

Stop alcoholism! Our families scream for help!

Drinking alcohol was always a part of traditional rural life in Russia; rural communities endorsed and participated in noisy and drunken festivities for various holidays, including religious celebrations. Nonetheless, many villagers condemned drunkenness if it sabotaged the wellbeing of a family, but they simultaneously treated someone who never drank with caution and suspicion.

Holidays were eagerly anticipated, and special occasions were always a time to eat and drink in abundance. The tradition was rooted in culture that seemed as ancient as time, and was reinforced by the claims that the heavy toll of peasants and their hard labor justified their drinking. It was also a way to join in the camaraderie of communal life and create a sense of belonging. It was the duty of all villagers to attend each family's celebration and share feasts, and exclusion from such events was treated as a sign of ostracism. But to celebrate always meant to drink.

All Russian people, including peasants in a largely-agricultural society, spent a large proportion of their income on spirits. Thus, by the late nineteenth century one-third of the state budget consisted of revenues from the sale of liquor and wine, and by 1913 over 10 percent of an average income was spent on alcoholic drink.[1] In addition to state-regulated sales, both legal and illegal production and consumption of alcoholic beverages coexisted in the country-side. Many rural centers had Russian versions of pubs (*kabaki*), and many widows and better-off peasants substituted their incomes with the profits made on the sale of homemade vodka (*samogon*).

However, the consumption of alcohol in the pre-revolutionary countryside was lower than among urban residents. In the early 1900s an average peasant drank 1.2 buckets of vodka a year, while an average urban dweller consumed four buckets of the said product (at about 8 liters per bucket, or 2.54–8.45 gallons). Rural women were not even considered sufficiently represented among alcoholics to be counted. For most rural residents, drinking was a special occasion, and among the rural population, on average 10 percent of peasants did not consume any alcohol while only 2 to 3 percent drank excessively.[2]

Hence historically, the image of a constantly drunk and perpetually drinking Russian village was a myth! Different proverbs and popular sayings in the countryside clearly reflected these notions, especially when they claimed that "if you drink to see the bottom, you will never see any good" or "if you drink too much wine, you will lose your brains." Unrestrained consumption was treated as a serious offense and almost a sin. Rural Russians, like most peasants elsewhere, knew that only hard labor produced solid harvests, and hard labor was only possible when one was sober. The village commune was also so closely knit that the moral upkeep of order in the village was considered a collective responsibility. Often, if a man went around the village and acted unreasonably under the influence of alcohol, the village elders or provincial (*volost'*) judges issued orders to whip the debaucher in public "as an example to threaten the others."[3]

In the early twentieth century, there even emerged a movement to foster sobriety and a complete renunciation of drinking in the countryside. Thousands of rural communities collectively renounced drinking, or, as many people phrased it, gave solemn vows to not drink at all. The Russian Orthodox Church actively supported such non-drinking communities. In 1914 in Kursk *guberniia* there were 114 of such communities in the countryside, compared to 15 in the urban centers of the same region.

Most drinking was limited to times of religious festivities, including Christmas, Easter, *maslenitsa* (a butter week, or seventh week before Easter), the Feast Day of Archangel Michael, and others. These traditions persisted well into the Soviet times, even if these days were not publicly pronounced as religious festivities. Most peasants went from house to house, and the celebration continued for two or three days, especially when family celebrations such as weddings were incorporated into the festivity. Because drinks and accompanying food were scarce and expensive, most peasants preferred to merge family matters with religious festivities, especially in cases of engagements and marriages. The popular saying about a father who "drank his daughter off" implied that the arrangements for a future marriage were brokered and sealed during one of such festivities.[4] Equally important were days when young men were seen off to the army, or when someone returned back to the village after a prolonged absence. Yet none of these celebrations were for the sake of drinking per se; they did not indicate the inclination to drink, but rather a tradition of merriness that persisted for generations.

Peasants never had their celebrations during high seasons of plowing and harvesting, and they avoided liquor during workdays. Moreover, peasants obeyed ordinary fasting times (all Wednesdays and Fridays) and the Great Lent that lasted for 40 days before Easter, when any intake of alcohol was considered a sin.[5] In addition, a number of other fasting occasions affected drinking patterns, including the so-called 12 great fasts that were observed throughout the year and included, in addition to the Easter Lent, the Fast of the Apostles, which lasted one to six weeks; the Fast of the Repose of the Virgin Mary; and the Christmas Fast, which lasted for six weeks before

Christmas. In addition, many social gatherings did not feature drinks as a prominent part; instead, women entertained the crowds by singing folk songs and dancing, and most common were *chastushki* (folk satirical rhymes) that were accompanied by *balalaika* or *garmon'* (a Russian small button accordion) played by men. In this setting, even men who were prone to abuse alcohol on other occasions often obeyed the demands of their spouses to restrain from heavy drinking.

Finally, the anti-drinking law of 1914 also restrained what the peasants spent on special occasions, thus leaving enough cash to improve their purchasing power for items such as grain and cattle. The cost of an average wedding dropped some 10 to 15 times as a consequence of this decree, thus once again freeing some scarce and much-needed cash for other items.[6]

However, the Revolution of 1917 tore through the fabric of an everyday rural life. A massive migration to urban centers displaced many previously rural residents, and even those who remained in the countryside started to imitate and mimic urban experiences and lifestyles that were not typical for a Russian village.[7] The village community started to catch up with the city, not only in its literacy rates or party membership, but also in how much it drank. The displacement of the ROC and the persecutions of its clergy also affected the drinking problem negatively. Although some rural residents continued to observe the Lent, most others felt unrestrained by the Church in their drinking patterns and went on to celebrate both traditional Orthodox festivities and new Soviet holidays by drinking heavily.

Peasant women became the most outspoken opponents of the new countryside lifestyle. One woman, for example, companied about the Soviet villages' failure to have "decent" Soviet holidays and celebrations:

> We now have too many holidays. There is something every weekend: the Day of the Construction Worker, the Day of the Metallurgy Worker, the Day of the Farm Worker, or the Day of the Retail Trade Worker. And since we now celebrate all holidays the same way, it basically comes down to communal or private (family) drunkenness. People spend their holidays outrageously poorly; [the holidays] are plain and boring. There is no tradition attached to any of them. Previously, each holiday was different, with its own traditions: for *maslenitsa*, [there were] pancakes and sledding; for Christmas, [there were] masquerades; for the Trinity, [there was] the making of flower wreaths.[8]

Because the 1930s and 1940s were tumultuous for the Soviet Union, the question of alcohol consumption was pushed to the margins on the list of social, economic, and political problems in Soviet society. But by the 1960s, the Soviet countryside emerged from this era of hardships with a different face—one deeply troubled by its addiction to strong spirits. The hardships and displacement that the people endured during the 1930s and 1940s also forced many Soviet people to rely on alcohol as a source of their solace. By the 1960s, even the public outcry against drunkenness became widely voiced:

> I just have to speak up my mind on the harm done to our people by alcoholic beverages. Everyone knows how harmful is their intake, yet the government continues to produce and sell them. They are not needed at all, because as medical research confirmed they are harmful no matter what. They are also harmful because the [work] discipline suffers as a result; people lose their dignity, their self-respect, and their talent. Relatives and loved ones become rude and harsh to each other. Children lose their parents to alcoholism, and then [the government has] to pay child support, for hospital stays, sick leaves, or even for an early retirement. I've heard once on the radio that the government does not want to produce alcohol any more but it fears that [should it do so] people would resort to drinking [cheap Soviet] cologne and other stuff and will get poisoned. But they drink it even now. And if someone gets poisoned, so be it, for the sake of saving other lives. The weeds should be gone from the field. Then the entire generation would not be poisoned instead.[9]

The Soviet system of centralized economy also transformed a traditional system of *magarych* that long since existed in pre-revolutionary Russia and was later borrowed and adapted to the Soviet life. Typically, large-scale domestic projects—building barns, fixing houses, some land work—were done with the help of neighbors and friends who received a payment known as *magarych*. On most occasions, it was a mutual support system that involved no cash transactions, at the very least because it was so difficult to assign cash value to any single project. *Magarych* was most often paid in kind, in the form of domestic products and progressively more often over time, of spirits. *Magarych* became almost exclusively about vodka or homemade samogon by the time rural women started to cry out against it in the 1960s. Most men even consumed it with those who provided it in the first place, thus turning this system of neighborly support into an occasion and an excuse to get drunk. One woman complained:

> The whole set up of our life forces men to drink hard. Collective farm-workers constantly need help from each other, and nowadays men go to turn the soil or help with the harvest only with a promise of a bottle of vodka. Collective farm directors are at fault for this. They should not push us into this, and should help out as needed on a more collective basis.[10]

The state-guaranteed employment, reinforced by male camaraderie, made firing debauchers and alcoholics from their jobs nearly impossible, especially in the countryside that constantly needed more hands and existed in an enclosed communal environment. All too often women took it upon themselves to publicly complain about alcoholism in the countryside:

We work and work hard. All women work with enthusiasm, [though we] have large families, four or more children each. But if we work hard, we also need to have a break, and this is something that we do not have. We spend all day at work, then go home where we do chores but also are supposed to rest. But instead we only see drunken rages of our husbands, their debauchery and hooliganism. We are forced to run away from the house and wait it out at our neighbors' or on the street in order for our [husbands] to calm down. [There is] alcoholism during public works [*obshchestvennye raboty*], at the milking farms, right in the political agitation centers. Instead of any political agitation work we only see half a liter of vodka on the table. Laborers and drivers drink together with collective farm chairmen. And we hear only profanities. Neither our agronomist nor our chairman pay any attention to it, it just had become a ritual, a custom of the sorts.[11]

The fact that there was little in the way of legalized persecution and restrain on alcoholics forced many women's organizations to appeal to the government for public recognition of the problem. The goal was to make either the government or local officials propose specific legislative and practical changes to curb the abuse of liquor by these women's spouses. In 1965, one such organization from Kursk region proposed the following in their letter to *Rabotnitsa*, the leading women's magazine in the country:

Please, help stop alcoholism! Our families scream for help! Drunks beat their wives and children nearly to death! They know no restrain. Why if a man hits a stranger, he goes to court, but if he hits a wife, threatens to kill her with a knife, scares the children daily and makes them emotionally unstable—no one persecutes him. We demand:

1. To forcefully remove such [husbands] from their homes, force them to work and to send a part of their pay to their families
2. Give women a right to divorce alcoholics free of charge and without a husband's presence or consent
3. If a family lives in their own house, take all rights away from a man and leave the house to a woman who raises the children
4. The same should go for flats; get rid of debauchers and transfer their right to housing to their families or their parents.

We are sure that only a harsh law will stop these drunken rages. We beg you, help us soon; the sooner the better; stop this nightmare—alcoholism![12]

All too often drinking went hand in hand with violence, and that is why medical professionals, lawyers, teachers, policemen, and even children of alcoholics appealed to the state to "take measures" to address a problem and create an effective legislative change that would allow the prosecution of routine alcohol abuse. They argued that:

> the most radical measure to liquidate the drunken crime is to pass a law
> that would forbid anyone from appearing in public while under the
> influence of alcohol . . . even if such laws would affect the sales of liquor,
> the losses of state revenues are incomparable to the losses of human life
> that occurs as a result of alcoholism.[13]

The Soviet system had some room for persecuting persistent offenders.
Specifically, it outlined regulations on persecuting the so-called slackers,
parasites, and antisocial elements. The definition was broad enough to include
various categories of people, but most common "offenders" were alcoholics
and drug addicts. Those charged with being "parasites" were often deported
to and resettled in remote regions of Siberia, starting with the eastern reaches
of the Omsk province. But the administration of various collective farms in
places of resettlements complained bitterly:

> only drunks, alcoholics and drug addicts are sent here under the pretense
> that they are parasites. But they continue on with their habits here and
> even demoralize their neighbors . . . there is only one way for them to
> get out of here, that is, to go to prison, but even this does not scare them.[14]

Moreover, there were two significant obstacles to *systematically* applying this
law to alcoholics. First, many men were the so-called functional alcoholics,
who went to work daily and did their job, and only got drunk and violent
once the workday was over and they got home. One woman wrote:

> It is common to think that if a man goes to work then everything is alright,
> he is not yet an alcoholic. But this is not true. All too often—and I can
> offer many examples—a man is a good worker, a good sailor or a coal
> miner, but once he gets ashore or to the surface, that man is gone. What
> is left is a selfish, annoying and vulgar drunk. In 1937–38 the personality
> cult left many of us widows, but now alcoholism is no less potent at
> making us widows as well.[15]

Second, many women did not want to take radical measures against their
husbands and were willing to put up with anything just "to keep a man in the
house." In the post-WWII rural Russia, men were a scarce commodity, and
most of them had a drink too many anyway. So for many women, to have a
family and a husband meant achieving the envious status of a married woman,
and this alone was a strong enough incentive for many women to remain
married to men who abused alcohol or other substances. When a drunken
rage was over and the crisis turned to a reconciliation, all too often women
sought not a divorce but someone to blame for their troubles, be that "unruly
friends," or "that woman who wants to whisk him away," or kiosks that sell
spirits, or anything else. These women argued both that "he just had no
choice" *and* "I just have no choice" and continued to live with their husbands
without any regrets or remorse, at least until the next drunken debauchery.

Children's complaints were also rarely taken into consideration. Some children became outspoken critics of their fathers' and step-fathers' drinking habits and appealed to local and centralized journals and magazines in a desperate outcry for help. "We have no life because of the father who is an alcoholic," wrote sisters Larisa and Olga.

> He drinks and fights every day, then throws knives at us, [or] an ax, or scissors. He spends all our money on drinks. Our mother is good but he humiliates her, threatens to kill her, and us as well. But our mother is too scared to complain about it . . . It is impossible to live like that anymore. He is a monster, not a human being; we don't want such father! Help us, please.[16]

In this particular case, both girls left the house when they turned 18 but their mother continued to live with the same man and the same set of alcohol-related problems.

If drinking was not accompanied by violence, then such households were even considered stable and functional. The argument in favor of preserving such unions included the discussion that drinking was not systematic but occasional. At least a portion of a man's wage ended up in the hands of his wife to be used for household needs, and at least occasionally he was available to give a hand around the house. It was impossible to *approve* drinking but it turned out to be possible to *justify* it, as one woman so candidly summarized:

> Some women say about their husbands: "Oh, he is so bad, he drinks heavily, and I'll divorce him." But I think this way. Yes, he drinks; yes, there are problems in the family because of this. But so what! He is the man in the house. And then he does not drink *all* the time. He gets sober at least on some occasions! At times he even brings his wages [home]. So don't waste your men away; they can be useful as well.[17]

Mothers wanted to safeguard their children from violence at the hands of their fathers and other male relatives. At the same time, these women were concerned that a popular acceptance of drinking and the wide availability of spirits might prompt their teenage sons to abuse alcohol as well. One mother complained that her son, aged 18, could not walk home from work without getting an offer of a drink.[18] Another mother emphasized that children know nothing and learn nothing about the harm of liquor, but they do see that their own and their relatives' drunken violence and fist fights go unpunished.[19] More than once women voiced their concerns that vodka was all too ready available. For example, a vocational school for the working youths had a vodka kiosk right on its premises! And many students got a drink or two between their classes. In this particular case, mothers appealed to the local administration to remove the vodka selling facility from the school, but to no avail.[20]

By the 1980s, it became obvious that drinking and alcoholism had reached epidemic proportions. Cases of alcoholism were even reported among rural women, which was a new trend.[21] Cases of complete alcoholism when all family members abused alcohol became commonplace as well. N.I. Ryzhkov, Chairman of the Soviet of Ministers of the USSR, wrote about the mid-1980s: "The country was really ruining itself by drinking. People were drinking everywhere: before a workday; after a workday; instead of a workday; in party cells and regional organizations; at construction sites and factories; in offices and at home. Everywhere."[22]

When Gorbachev initiated his anti-drinking campaign in 1985, he argued that "the initiative [to start the campaign] belonged to the people, really. The party organs routinely received avalanches of letters from mostly wives and mothers about it. They described horrible examples of family tragedies . . . and drunken crimes."[23] The Central Committee of the Communist Party then issued a resolution, dated May 1985, "On Measures to Overcome Drunkenness and Alcoholism." At its basic level, the resolution cut the sale of liquor in half in 1985–86, while the prices of liquor went up by the factor of two.[24] But the measures proved too radical and rushed; more than anything, they undermined state revenues and subsequently the state budget. In 1985–88, the state budget came 67 billion rubles short of anticipated revenues, and the anti-drinking campaign had to be relaxed. As a result, in 1989 the revenues on the sale of liquor exceeded similar revenues from 1984, thus replenishing the suffering state budget but also undermining any previous attempts to minimize alcohol consumption.[25] Whatever the shortcomings of the anti-drinking campaign, to the present day most Russian citizens, and especially women, consider battling alcoholism one of the most crucial concerns that has to be addressed systematically.[26]

In the countryside, the campaign had a different dimension compared to the largely urban Soviet Union of the 1980s. Although the regulations specifically outlined that liquor had to be sold only after 11:00 am and only to people over the age of 18, most retail workers in the countryside felt that they could do endless "favors" to fellow villagers by curtailing and bypassing those regulations. As a result, it was possible to buy liquor at any time of the day, and even on credit (i.e. in exchange for a promise "to pay for it soon")! One woman complained sarcastically that

> Our village has only one store and it is headed [i.e. the director] by E.I. Savinova. Oh, what a kind soul! Even if there is a queue for bread or anything else, once a half-drunken man approaches her, she leaves everything behind to satisfy his cravings. And if any man has no money or does not have enough, she will even write it off as his debt in that big black book that she keeps hidden under the counter![27]

Many families only learned of the new unofficial credit system when the husbands or fathers received their pay but brought next to nothing home; instead, they left most of it in the liquor store to repay the debts.[28]

In the countryside it was also usual to consume homemade wines or samogon (homemade vodka). In the 1960s and 1970s, at least half of all liquor consumed in the countryside was homemade. The annual consumption per capita was estimated at about 2 liters of pure spirits.[29] And the villagers did not make a secret of it. They honestly admitted that "at least every third person moonshines" or "it is a rare house where no one produces samogon."[30] It was also considered a woman's responsibility to make samogon.

The government was fully aware of the problem, especially the increase of moonshining in the face of more restricted sales of liquor in the second half of the 1980s. In 1987, the Politburo insisted that harsher measures be taken against those who made their own liquor. Specifically, as of October 6, 1987, the Politburo adopted a measure to confiscate property in cases of the most excessive [*zlostnyi*] samogon making. But on most occasions, people who were prosecuted on charges of illegal trade in homemade liquor were not subjected to the confiscation of property.[31] As a consequence of this lax application of law, in 1990 about 1.5 billion liters of samogon, which would be equal to about 75 percent of legitimate sales of all spirits in the country, were produced by people in their homes. The bulk of distilling took place in the countryside, as on average urban residents made only 2.3 liters of wines and other alcoholic beverages at home, while rural residents made 22 liters.[32]

With the introduction of the anti-drinking campaign, complaints from rural residents about the abuse of samogon also multiplied.[33] Most prosecutions resulted in fines and the confiscation of distilling equipment, and only few cases were considered criminal offenses. In 1985, there were 45,000 registered cases of producing, selling, or possessing homemade alcohol. The number went up to 122,000 in 1986. But most of these cases were resolved with the payment of fines. In the period 1985–89, only 13,000 people were prosecuted for moonshining as a criminal offense.[34] Because samogon production was a woman's task in the countryside, over 70 percent of those arrested and charged with bootlegging were women.[35]

These women, however, did not necessarily drink what they produced. Female alcoholism has always been an exception rather than a rule in the countryside. Many women feared that if they started drinking they would damage the family's reputation in the closely-knit rural community, while others were too preoccupied with daily chores to take to drinking seriously. But even so, the number of female alcoholics slowly started to rise by the end of the Soviet era and into the post-Soviet Russian life. According to medical practitioners, the number of female alcoholics doubled between 1930 and 1980, by some estimates reaching 10–15 percent of the total number of alcoholics (on occasion numbers as high as 30 percent are also cited).[36] Most women alcoholics develop an addiction when they are 25–45 years old, while most males become addicted before they turn 25. Women also develop dependency faster than men; while it takes 3 to 7 years for a man to create a stable and consistent dependency, for women it takes only 1 to 3 years. Most women who drink regularly and consistently for 5 years become fully dependent, whereas only one-fifth of all men develop similar dependency in the 5-year time span.[37]

Women and men also drink for different reasons. According to research in psychology, women have a lower level of resilience when it comes to family problems and private lives, and many women fall prey to the drinking habits of their male partners. And female alcoholics also behave differently. Women mostly drink at home by themselves, many attempt to hide the fact they are drinking, and relatively few commit drunken crime, of which most (40 percent) relate to drunk driving. Furthermore, women-drunkards are less likely to lose their job, less likely to drink before noon, but more likely to commit suicide than men. Often drinking is triggered by illness, which only grows graver as drinking becomes more abusive. But male partners still play a crucial role. Women who experience a long time living with husbands who are alcoholics at times learn to cope and function best in that setting and know no other type of family relationship. As a consequence, they enter a vicious cycle, i.e. when they break with one partner, they subconsciously choose men with similar drinking problems. Unable to handle the stress of repeated problematic relationships, such women join their partners and start drinking themselves. But considering that female alcoholism is less prone to successful treatment, and considering that most women prefer to hide the fact of their drinking, such women often damage their health and create addictions that are irreparable or hard to treat.[38]

Female alcoholism also has a specific social profile. Of women who developed alcohol dependency in the Soviet Union during the 1980s, 24 percent worked in factories or were involved in the construction industry; 49 percent were employed in the retail trade and fast food industries; and only 2 to 5 percent were white collar workers, specifically teachers, medical professionals, and artistic and scientific elites. Most of these women preferred hard liquor over other drinks.[39] In 1988 alone over 267,000 women were taken into custody by the police for drunken behavior in public; and over 21,000 women were arrested for begging or being "bums." The number of women who underwent mandatory treatment in rehabilitation centers increased by the factor of two from 1985 to 1988, reaching an astounding number of 17,000 women by 1988.[40]

This near-epidemic of alcoholism reached unprecedented proportions in the 1990s. In only the first few years the number of mental disorders that are a consequence of alcohol over-consumption went up by a factor of five. Various studies estimate that in the first half of the 1990s, at least 10 percent of all males and 3 to 5 percent of all females were suffering from addiction to alcohol. Statistically, the consumption of spirits in the Russian Federation at the time exceeded consumption in any other country in the world. In 1990–95, an average intake of pure spirits (not by volume but by alcohol content) was 18 liters annually in Russia, compared to 12 liters in France and the safe guideline for consumption, which is 8 liters (the level of 1989 consumption in the USSR). But what made the difference was not the amount of spirits per se but the quality of the spirits. Most drinks that are consumed are high in alcohol content (vodka over wine, for example) and the quality is appalling; 60 percent of all liquor sold in Russia at the time "did not stand to any criticism."[41]

Thus it is unsurprising that the peak of deaths from poisoning with alcoholic beverages was in 1994, when 56,240 people perished as a result of their intake of low quality spirits. This number was twice that of 1992, and five times the number in 1987. In addition, another 68,800 died of illnesses associated with an excessive intake of alcohol, including cirrhosis of liver and alcoholic hepatitis. The majority of these people were of working age. According to various estimates about a fifth of these deaths were women. If one includes all deaths related to drunkenness (accidents, acts of violence, illnesses, poisoning, and others), the number of total human loss to liquor in Russia was estimated between 400,000 and 700,000 people in 1994 alone![42]

To the present, many of the problems in rural Russia, including depopulation and lack of a sufficient workforce, relate to alcoholism. Some women even complained that:

> It was better after the war [World War II] because we had both, a work routine and holidays, and also because we believed in a brighter future. But now on average each person drinks 15 liters of vodka. Families fall apart because of drunkenness, sons kill their fathers, [and] women give birth to disabled children [who are] invalids.[43]

There is no limit on the sale of liquor in the post-Soviet Russia, and the price of vodka is on average 40 rubles per liter.[44] As a result, a bottle of vodka in

Table 12.1 Reasons for drinking (% of all respondents)

Reason	Men	Women
Broke up with my lover	0.1	–
Life sucks	–	1
I can't handle being alone any more	0.4	0.4
Poor living conditions	0.5	–
Bad weather	0.5	–
Because of illness	0.5	0.4
I am desperate	0.6	1.6
Had a fight with a spouse	1.0	–
I am sad	1.1	3.5
Others were drinking	3.3	2.4
Personal special occasions	3.4	21.9
I had a bad hangover	4.6	–
Because it's a weekend/holiday	6.2	3.2
It's a payday	9.1	6.0
Meeting with friends	19.8	3.1
I just wanted a drink	37.4	46.7

Source: I.R. Takala, *Veselie Rusi. Istoriia alkogol'noi problemy v Rossii* (St-Petersburg, 2002), p. 262

some regions such as Pskov costs less than a liter of milk. New rehabilitation services for treating alcoholics were introduced as well, the most popular one being the use of hypnosis. Recent opinion polls further demonstrated that in 2005, 60 percent of residents of Russia believed that another anti-drinking campaign might benefit Russian people, and an equal number argued that everyone started to drink more after the demise of the Soviet Union than they had done before. Most supporters of another anti-drinking campaign were women.[45]

13 The female face of the criminal world

Honor once lost never returns.

Russian rural life before the October Revolution was rooted in the Orthodox tradition, and thus made a clear distinction between a sin and a crime. That distinction did not necessarily follow the criminal code or support the opinions and practices of the judges and criminal prosecutors. But a Russian village continued to function according to its own norms and shied away from prosecuting for sins that were not seen as crimes. The label of a sin was used for everything that went against God, namely work during religious holidays, the use of profanities, petty theft, and a child born out of wedlock. Crime, on the other hand, included murder, major theft, robbery, and some fist fights. But every deed in the countryside was first assigned a moral judgment. The moral verdict of the entire village community was often the most respected and feared judgment. Regardless of the criminal code, or even the severity of the crime and legal punishment, the community could justify and explain away even a murder when the community saw a legitimate explanation for it, or could condemn a person to ostracism for a sin that was not punished by law. This community judgment was final and irrevocable, and many rural residents feared such judgment more than any legal sanctions. The reputation of the family was hard to maintain but easy to lose, and condemnable stories could taint such a reputation for generations to come.

But the suffering and transformations that the first two decades of the Soviet regime brought to the countryside tore through the well-established fabric of rural life. The Soviet Criminal Code introduced new categories of crime, such as political crimes, and women and men were often prosecuted at the whim of the state. Collectivization, the famines of the 1930s, and various repressions disproportionately affected rural life (rather than city centers), partly because most of the Soviet Union was still rural at the time, and partly because many measures were explicitly aimed at the rural population. The Harvest of Sorrow famine of 1932–33 alone took at least 8 million rural lives and brought a massive human tragedy that the famous Russian writer, Mikhail Sholokhov, claimed "impossible to erase from the memory till death."[1] The tragedy of dekulakization was such that even most ardent Komsomol (Youth Communist League) activists on occasion could not withstand the horror of the campaign and went clinically insane from encountering human misery first-hand.[2] By dividing the entire population into first (most dangerous) and second (less dangerous) categories of kulaks (wealthy peasants) and enemies of the people, Stalin's government "cleansed" the countryside of 400,000 people of the first category of kulaks, most of whom were executed in the 1930s.[3] In the 1930s specifically, and also on a lesser scale until the 1950s, deportations and resettlements directly affected nearly 6 million people, of which at least 2.5 million were resettled in the so-called kulak resettlement campaign of the early 1930s.[4]

The experience of deportation was often divided along the gender lines, but the resettlement affected both men and women alike.[5] Kulaks were sent to a special settlement, along with their family, meaning a wife and children, first up to the age of 10 and later to age 14. All their property and possessions were transferred to the collective farm. But in practice, many of the neighbors did not even wait until the deportees left to pillage their valuables. People went into their neighbors' houses to pilfer spoons and pillows, shoes and dresses. The scene of the deportation often turned into complete chaos, accompanied by the screams and tears of devastated women, with crying children at their sides. The deportation did not spare anyone, "neither old men nor women, nor pregnant women, not invalids on crutches." Although the deported kulaks were presumably wealthy, in reality they were "average Russian peasants, both men and women, in *zipuns*[6] over a long shirt and many even in *lapti* [Russian bast shoes]. Children of all ages were swarming in the midst of this chaos as well."[7] Those who survived the journey and the first year in their new place of residence were put in the category of *lishentsy*, or those without rights such as freedom of movement. They were legally disenfranchised, and continued to live like this for years to come.[8]

World War II brought new political repressions. There were some female criminals but many women were repressed because they were "wives of enemies of the people," and because they were the wives of "traitors of the motherland," a broad category that applied to nearly all POWs and some DPs after World War II. The spikes of political repressions against women were similar and parallel to overall dynamics of terror. Thus, these repressions reached their height in 1930 during the collectivization campaign and then again in 1937–38 during the Great Terror. Yet women also suffered massively during World War II when the proportion of women in the GULAG rose from 7 percent in 1941 to 26 percent by 1944.[9] Yet, by and large, the number of political repressions declined after the war, and prosecution and imprisonment for criminal activities became predominant over other types of charges.

The famine years of 1946–47 saw a peak in criminal activities among women. Most of those were hunger and desperation crimes, most commonly petty theft. This is why most women who were prosecuted in the immediate post-war period had no prior criminal background and lacked any means of livelihood. In one of many examples, in 1947 Mikoianovsk regional court of

Belgorod region presided over a case of a 47-year-old illiterate peasant woman, M.F. Kazakova, who was a widow and a mother of two children aged 5 and 19, for stealing 5 kilograms of sugar beet and 600 grams of rye ears. The woman was sentenced to 5 years of labor camps for theft without loss of property (simply because she had none), and her 5-year-old child was placed under the custody of her brother.[10]

In a desperate attempt to curb desperation crime and to protect its meager agricultural produce, the government cracked down hard on all those who stole even a single ear of grain. As a result, in 1946–49 nearly 50 percent of all new GULAG inmates were women with small children. As of July 1, 1947, 19,000 children under the age of 4 were with their mothers in the GULAG, and 6,800 women were sentenced and imprisoned while pregnant. The number of children born in the GULAG exceeded the capacity of GULAG orphanages by three times, and for that reason infants stayed with their mothers and other inmates in the camp barracks. Children over the age of 7 were not allowed to accompany their mothers, and if there were no relatives, or if relatives refused to take in an extra mouth to feed, these children were put into orphanages. As a result, in many places such as the Ivanovo region, of all children in orphanages in 1946, 17 percent were children whose mothers were in the GULAG system, and the number increased to one-third of all children in 1947.[11]

Most children born in the GULAG were undernourished, physically weak, improperly cared for, and required living conditions and hygienic standards that the prisons could not offer. This "children's problem" became so pronounced that in 1947 the government decided to extend an amnesty to 15,000 expecting mothers and mothers with children under the age of 4 years. But the amnesty was of little consequence for most peasant women. Women who were sentenced for treason, espionage, terrorism, banditry (all were common categories common during these days), and crucially, for theft of socialist property, were excluded from the amnesty. The definition of "socialist property" included grain ears and beets as all agricultural produce belonged to the state. And because most peasant women were sentenced during the famine for stealing food, they were largely not subject to amnesty, yet this amnesty was readily applied to professional pickpockets and women with extensive criminal backgrounds. Moreover, as the government was freeing some women with one hand, it was imprisoning ever more women for the theft of state property with the other hand.

When a woman was arrested the entire trial did not take long. If a written statement was received from a collective farm, the woman 'thief' was taken into custody, and most women immediately admitted their guilt. This paperwork was all it took for the court to sentence a woman to a labor camp. Most cases that went through the court system were identical. A collective farmworker, M.A. Marchenko, was sentenced to 5 years of labor camps for a theft of "rye ears of a total weight of 850 grams, which were discovered during the search," while two other women, Maslieva and Propashkina, were sentenced for 6 years of labor camps for the theft of 4 kilograms of potatoes.[12]

According to official data, by the end of 1948, 503,000 women were serving their sentences in various prisons and camps, including 9,300 pregnant women and 23,790 women with infants under the age of 4 years. In 1949, thanks to the amnesty, 56,000 expecting and new mothers were freed before completing their terms, and soon thereafter another 29,000 women followed.[13] Yet, as was the case with the 1947 amnesty, the amnesty of 1949 did not include those sentenced for the theft of state property, and women who served their time for desperately trying to get some food for their children, even if caught stealing 400 grams of rye ears, continued to serve their sentences.[14]

Female crime rates started to decline and leveled out in the 1950s with the relative normalization of everyday life. Of course, the pronounced building of a communist society implied an imbedded hope that crime would cease to exist. Yet even though this goal was utopian, the dream of "living under communism that has no room for crime" seemed rather real for many people. According to Soviet official statistics, nearly 100 percent of all criminal cases were solved, which boosted propaganda campaigns to curb crime. And although this number was without any doubt highly exaggerated, there were no anomalies or spikes in female crime up to the late 1980s and early 1990s.[15]

Once Russia entered the age of democratic reforms in 1990s, the crime rate went up dramatically, which according to the opinion of many lawyers, "is a real cost of people's freedom, which can be used not only for the acts of kindness but for evil as well." In 1992, the number of crimes in Russia went up by 45 percent, but the increase was not as dramatic in years to come.[16] Of those crimes, 58 percent were solved in 1990 but only 52 percent were solved in 1991. However, this number is compatible to an average of around 50 percent for most countries.[17]

In the early 1990s many people became disillusioned with the official system of prosecution and failed to even register crimes. This, of course, obscured the real rate of criminal activity in the early 1990s. For example, only 10–12 percent of people relied on the police and court system for protection, whereas an equal number of people believed only in protection extended by racketeers and other criminal gangs. Some people expected help and protection only from friends (26 percent) or relatives (25 percent). Many people were so scared by the early post-Soviet chaos and lawlessness that 36 percent of urban residents believed that there was no place to seek protection or justice *at all*, and 59 percent (!) of rural residents supported this view as well.[18] As a consequence, it has been estimated that only 1 in 10 criminals ever faced the judicial system in the 1990s, and only 2 to 5 percent were legally prosecuted and found guilty.[19]

But the general profile of crime was still highly gendered. Thus, of those who were charged with breaking customer protection laws (for example, cheating customers), 59 percent were women; of those charged with producing home-made samogon, 63 percent were women; and women made up one-third of all those who were charged with corruption and with exceeding responsibilities and powers vested in them by their jobs. In comparison, for all other crimes, including theft, manslaughter, and robbery, only 8 percent of perpetrators were female. Two-thirds of women charged with falsifying sales and/or cheating customers were rural women.[20] Ironically, 60 percent of all people in Russia at the time found it reasonable and justifiable to "cheat in small ways."[21]

Overall, the crime rate increased faster in rural Russia than in the cities (although in absolute numbers urban crime still exceeded rural rates). In 1992–98, the crime rate in the countryside went up 1.5 times, which was more than 11 percent faster than urban Russia. In 1990 only a quarter of all crimes were registered in villages, but the share of rural crime rose to one-third by 1998.[22] Many rural residents blamed the collapse of the economy for this rapid criminalization of the countryside:

> It is very hard to find a job. And if you find one, there is a risk of losing it any time . . . The youngsters do not have any place to work or any place to realize their knowledge and their ambitions. Hence they are as poor and desperate as the elderly of the village. It looks the following way here: none of the businesses work; everything falls apart; people lose their jobs daily due to layoffs; unemployment continues to rise; and youngsters are left with nothing but to do drugs, acts of vandalism, and petty theft.[23]

Typically, women received milder sentences than men. While 34 percent of all men who faced the court system were sentenced to imprisonment, only 17.7 percent of all women received a similar sentence. This difference, though, was due to the nature of the crimes rather than any explicit gender-based discrimination. Women were subject to fines rather than imprisonment more often than men. Of 731 prisons that exist in the Russian Federation, only 33 are for women (and none are mixed). But because the number of female prisons is low, women served their terms far away from home more often than men (50 percent of women compared to 20 percent of men). The great distance that separates female inmates from their homes often takes away these women's chances of seeing their children and families, even when visits are officially permitted.[24] About 30 percent of men and 40 percent of women sentenced to imprisonment have a right to leave their prison and have vacations of 12–18 days with their families. However, only 6 percent of women and 3 percent of men have the financial means to realize this right.[25]

According to the current Russian Criminal Code, the prison term can be postponed for women who are expecting or have infants under the age of 3 years. Yet not all women are entitled to this right, and hence of 33 female prisons, 10 have orphanages where children live while their mothers serve their terms. Nearly all mothers with infants are allowed some time at the beginning of their sentence to take their children to relatives or to orphanages of their choice. Unfortunately, recent research indicates that most women take that time and the monetary allowance provided to address their personal problems and to buy cigarettes and cosmetics instead of caring for their children.[26] All in all, the twenty-first century has so far been marked by the feminization of criminal activities, with a rapid increase in the proportion of female criminals.[27]

14　Women of the oldest profession

Honor buys no meat in the market.

It is noteworthy that post-Soviet Russia saw a "new" problem that *officially* did not exist in the Soviet Union, namely prostitution. Up until the mid-1980s, only occasional police or media reports mentioned "women of amoral behavior" but always did so with an implied understanding that such women were an exception rather than a reality of Soviet life. Because of the complete denial that prostitution existed in the Soviet Union, very little reliable data exist of this phenomenon during the Soviet days. It was only in the late 1980s that the so-called "night butterflies" received attention from first writers, journalists and filmmakers, and then from lawyers, social scientists, medical professionals, and finally law enforcement officials.

Prior to the Revolution of 1917, prostitution was not considered a criminal offense. Ever since the times of *Sobornoe Ulozhenie* (the Legal Code) of Tsar Aleksei in 1649, "flesh mongers who forgot the fear of God and the Christian Law" were "to be beaten with leashes," but women were only verbally condemned for their behavior. Moreover, prostitution existed only in urban centers; people of the countryside did not know of it as a profession. At times, wives of soldiers were somewhat an exception to this rule. Ever since the early eighteenth century, soldiers served in the Russian army either for life or, later, for 25 years. So it was likely for many women who married at 17–18, and saw their husbands off to serve in the army, that they would never see their husbands again. Such wives often "had liaisons with strangers and other men." Yet the village mentality did not condemn such behavior and was rather indifferent to it, and even "children born by soldiers' wives out of wedlock were to have equal rights as all legitimate children."[1] Some villages had women "who were all too available," according to the village morale, but such women were the exceptions and were well known in their communities, and rarely received regular payments for their sexual favors.

To the present, despite the changes in legislation and lifestyles, only managers of brothels and pimps are prosecuted according to criminal law, while prostitutes face administrative charges. By the late 1980s small fines levied by the state made this seemingly easy and profitable occupation attractive to some women. For the first time in the history of the Soviet Union, the government admitted that this profession existed in the USSR, and some 2,000 prostitutes in Moscow in 1988 had files on them in the law enforcement offices. The number increased to over 5,000 in 1989 for Moscow alone, and it was estimated that 20 percent of the "newcomers" whose filed were added were 12–13 years old.[2]

This number did not reflect the full scale of prostitution, even for Moscow, because it only reflected those women who were arrested and whose profiles were recorded in official databases. Various researchers have estimated that in Moscow alone at least 10,000 women made their living in this profession in 1988, and similar increases were evident in nearly all regions of the Soviet Union. Social scientists argued that economic and the political transformations of the country were among the main causes for the rise in prostitution, although equally important were factors such as the influence of the media and press.[3]

At first, most media reports on prostitution were sensational and tended to glorify and romanticize the experiences of prostitutes. Most journalists and authors looked at "hard currency prostitutes" (those working for foreigners) and, with a clearly identifiable sense of envy, portrayed their lifestyles, their incomes and their access to highly desirable foreign goods as almost worthy of admiration and imitation. This image had a negative impact on the younger generation, as some teenagers looked to these women as role models. Several social research teams distributed questionnaires among high school students in St Petersburg (then Leningrad) and Riga in 1988, asking them to identify their future profession. Shockingly, the profession of "hard currency prostitutes" made it into the top 10 most desirable, most prestige and most profitable professions that these girls aspired to have.[4] Furthermore, in 1987 only 1 in 20 prostitutes were under 18, but by 1996 1 in 8 sex workers in Russia, or 1 in 5 in Moscow and St Petersburg, were not yet of legal age.[5]

In the late 1980s prostitutes were assigned to two general groups: the so-called "train station prostitutes" (low profile, mostly working with low-income males) and "hard currency prostitutes" (elite escort services, mostly for high-income men and foreigners). Among high-profile sex workers of the late 1980s, 70 percent were under the age of 30, whereas 68 percent of low-profile prostitutes were 30 years old or older. Most elite prostitutes were age 18–30 (14 percent in the 18–20 age group; 30 percent in the 21–25 age group; and 20 percent were 26–30-year-olds). Comparatively, the largest group of low-profile prostitutes (35 percent) was much older, falling in between the ages of 41 and 45. Of elite prostitutes, 65 percent had a high school diploma and 23 percent had college degrees, while most low-profile sex workers lacked even a high school diploma.[6]

Interviews with some prostitutes also revealed a glorified and romanticized vision of the profession before entering it. Thus, one woman said:

We were lied to all the time! [We were told] one thing at school, another at home. Why did I get into this? Did I dream of *this*? No, I dreamt like we all did about all kinds of naïve nonsense. And I was mad that my female classmate had [all her clothing and stuff] from abroad while I only had stuff for Soviet mass consumption. [This was so] because my father lived an honest life while her daddy was into some illicit business. Also, our teacher loved to get gifts from that girl and surely made her special among us, girls who were simple and had nothing. Hence I wanted everything to be fair, and I started to go around.[7]

The lack of law enforcement made prostitution appear even more attractive to some women. In 1887–90, of 1,327 prostitutes arrested in Moscow, only 18 were taken to court on administrative charges and fined for their activities, and 29 more were fined for spreading STDs and refusing to get any medical treatment. Equally problematic was the spread of pornographic materials since no legislation existed to prosecute even those producing child pornography.[8] It is unsurprising that shortly after the Soviet Union collapsed, the majority of both men and women approved of introducing state censorship to control and monitor the production and sale of sexually offensive materials, as well as some censorship for sexual content in the mass media.[9]

Almost all activities of elite prostitutes in the 1990s were controlled by pimps, racketeers, and criminal gangs. Of elite prostitutes, 80 percent admitted their full dependency on "managers," whereas 93 percent of women who worked in upscale brothels said the same. Only low profile "train station" workers reported a low level (15 percent) of using pimps to finding them clients and managing customer relations and payments.[10]

In the 1990s, Russian prostitution also expanded abroad. Although legitimate labor migrants and mail-order brides were the most common groups of women among those who travelled abroad, some women ended up working as prostitutes. Officially, in the 1990s, 50,000 women left Russia annually for permanent residence abroad, and up to half a million Russian women traveled abroad temporarily to work for 1 to 2 years. The main goal of 86 percent of women who left Russia, both temporarily and permanently, was to make more money.[11] Moreover, of those who left temporarily, about a half hoped to change their immigration status and to stay abroad permanently, and about 20 percent hoped to find a man and stay abroad through marriage.

The average age of women who leave Russia is 26, and 1 in 10 is younger than 20 (comparatively, the average age for men who go abroad is 35). A third of all women who leave Russia have never been married, and another third are divorced. Many women hope to start off as domestic workers (house cleaning, childcare). Yet there are very few legal channels for low-paid employment abroad, and therefore many women travel, presumably for tourism, and end up becoming "a risk group." Because there is no legal protection for women who violate visa regulations and have no legal status in a country, an estimated half a million women per year are trafficked as sex workers to Western European countries. In some countries, such as Germany, an estimated 60–80 percent of all women sex workers are from Eastern Europe and the newly independent countries of the former Soviet Union.[12] Although some women recognized the risks of their travel abroad to seek employment without documents and legal rights, many women aspired to a more legitimate employment and were forced into the sex industry either to mitigate harsh financial circumstances or by the deceitful and illegal actions of sex traffickers.

15 Religion

We only want to pray to God and praise our Fatherland.

Some newly acquired territories and regions on the outskirts of the Russian Empire practiced other religions, but traditionally Russian villages were predominantly Orthodox. Even for Russia as a whole, by the early twentieth century 103.4 million people, or 65 percent of the entire population, considered themselves Russian Orthodox.[1] Hence the pronounced changes of the Bolshevik Revolution that declared religion "the opium for the people" were not easy to implement. Many rural residents resented or even openly rebelled against attacks on their religious lives. Those who welcomed the change were predominantly males who were inspired by the promise to rid their loved ones of "religious oppression," and younger women among the Komsomol workers. For a country where over 80 percent of its population was rural, breaking old religious ways was not an easy task to achieve. The new Soviet government fought a fully fledged war on religion, and the main opposition came from rural women. It was not until the collectivization campaigns that the anti-religious propaganda started to bear some visible results, and even then the infamous *babyi bunty* took peasant women to the forefront of the opposition to the Soviet government.

The separation of the Church and the state was announced by the Decree of January 23, 1918, and from that moment both the most ardent supporters of the Church and clergymen suffered from unjustified repressions. By 1922, over 600 monasteries were closed down throughout Russia, and the number of laymen and clergymen persecuted by the Reds in 1917–21 exceeded 10,000 people. In 1922–23, 2,691 priests, 1,962 monks and 3,447 nuns were shot. In 1929–36, over 50,000 people, both ordained and laity, were persecuted for their religious beliefs, and 5,000 of them were shot. With the advent of the Great Terror, the number of those repressed on religious grounds reached 100,000 in 1937–38, although these numbers reduced during the early 1940s to some 2,000 people.[2]

The Constitution of 1936, in article 124, formally recognized freedom of "religious cults" with the equal "freedom to carry out anti-religious propaganda." It further formalized the separation of the Church and the state and prohibited any religious instruction in the school setting. Yet no matter what the Church suffered as an institution, popular religiosity remained strong. Even the official census of 1937 demonstrated that 55.3 million people aged 16 or older listed themselves as religious.[3] In addition, there were indisputably those who held on to their religious beliefs but feared declaring them on the census questionnaire. Two-thirds of all women declared themselves followers of an established religion, but more than a half of all men were self-proclaimed atheists.[4] Russian Orthodoxy was still the predominant religion, with 41.6 million officially counted adherents, and 70 percent of them were women.[5]

The government aimed to follow official resolutions when it came to closing churches down, i.e. the initiative had to belong to the people. The main document was usually a petition from a local party cell with a consensus of the entire village to close down the church, typically with a pretention to using the church building for other purposes or doing a new construction project in its place. Residents of Peria village in Vologda region petitioned to close down two churches in order to turn them into a school and a club for collective farm youngsters. Some ardent believers opposed the petition, but to no avail. The promise of a school never materialized, and church buildings were "sold" as scrap building materials; dynamite was placed under both churches and leftover pieces were collected by local builders. Both churches were architecturally significant and dated back to the late eighteenth century.[6]

Church books and archives were treated as recyclable paper; priests' robes were given to tailor shops to be used as cloth for uniforms; and icons, crosses, bells, and other metal and wooden objects were recycled, often broken or smelted, and used in repairing other buildings. Everything that could not be recycled or used for non-religious purposes was burnt, destroyed, and broken.[7] And villages responded immediately by producing many rhymes that reflected this tragic reality, singing that "the father makes samogon to drain his sorrow because his son used all icons to heat the house."

Such rhymes also reflected the gender and generational divisions that emerged in this anti-religious campaign. The campaign itself was more successful in destroying the visual attributes of religion than in destroying the faith. Most active participants in their fight against "opium for the people" were young men and women, usually in their late teens, some professionals, and some males, but there were only a few older women among them. Peasant women became the stronghold of religious life, and they often took it upon themselves to preserve the practices and morals of the Christian teaching. They were also the ones to rebel openly against the Soviet government's acts, and they fought, at times hard, for the preservation of Church property and possessions. In nearly all registered cases of protests against "excesses in the anti-religion campaign," the instigators and leaders were rural women. In some villages up to 500 "faithful peasants," armed with axes, shovels, and other household items, opposed the destruction of the local church.[8]

The resolution of April 2, 1929, "On Religion Organization" denied the Russian Orthodox Church the status of a legal party. All religious organizations and their members were supposed to register with the local authorities. All religious life was limited to services in the church, and the priests were required to turn in to the government 75 percent of all donations as a state tax.[9] The ROC also became a major source of income with a financial potential not to be missed. Confiscated Church possessions were important, but so was the new "anti-bell campaign."[10] Taking all bells off the churches and smelting them for recycling was a way to kill two birds with one stone. First, it attacked the very heart of Russian Orthodoxy, which was heavily centered on the bell-ringing culture. Second, it created a readily-available pool of scrap metal to be used for other needs. But, once again, the grandiose plans of restructuring the Soviet religious life were met by significant opposition in the countryside, especially among the peasant women. For example, news of the opposition to closing the church in Chernaia Zavod' village in Yaroslavl region reached as far as Moscow. In a note to Stalin, Lavrentii Beria reported on September 30, 1938, that the local party cell decided to close down the church and remove church bells from the bell tower.

At the same time a group of parishioners gathered in a crowd of about 500–600 people, who, standing by the church building, declared emphatically to the people who came to remove the bells: "we will not let you take down the bells, and if you would dare to start taking them down by force, we will find our revenge and you will meet our anger right there." [11] Every time a party cell representative attempted to approach the church, there were shouts in the crowd: "Bandits have come, let's beat the hell out of them," or "Help! We are being robbed!" The most active were women. One female collective farm worker, E.P. Baulina, "called from the church doorsteps for people to arm with axes and shovels and to make all offenders run away, and if they would persevere, then chop them all into pieces." Around the clock villagers took turns to monitor the actions of the party representatives and workers from the city. "Every time watchmen saw a car approaching the village, they started yelling, and all peasants dropped their field work and ran to the church." Somewhat paradoxically, Stalin ordered an investigation into the matter and as a consequence reprimanded local authorities.[12] The peasants of this village, and most of them were women, won the battle and saved the church.[13]

World War II brought the liberalization of religious policies, although somewhat limited in scope. The government aimed to enlist all possible means to boost the morale of the Soviet people in fighting the war, and the ROC stood ready to use the opportunity, appeal to the people, and improve its position vis-à-vis the state in the process. Financially, the ROC provided the means to build and equip the tank division named after Dmitriy Donskoi, and overall it transferred 300 million rubles to the state for the war effort.[14] But its main contribution was spiritual. With their men at the warfront, many women of the countryside turned to the Church for emotional support. The faith, hope, and boosting of morale that the ROC provided were invaluable to many people and the overall war effort. In 1943, the governmental announced "a new direction" in the state and ROC relationship, allowing the ROC to exist in a socialist space, even if under a direct supervision of the state.[15] In September 1943, the Soviet of People's Commissars approved the establishment of the Committee to oversee the needs of the ROC, and the main function of the committee was to oversee the relationship and regular communication between the Soviet government and the Moscow Patriarchy.[16] Members of the committee also served in the People's Commissariat of the State Security, and the Committee was under the direct and exclusive control of the state.[17]

In 1946, the ROC was granted the status of a legal party with limited rights, which nonetheless allowed the ROC to buy transportation and buildings, and to construct its own buildings. Its lay members, who serviced the church, again mostly women, were allowed to join labor unions and received social benefits equal to other workers.[18] Some of the changes were also evident in the number of open and functioning churches. Immediately prior to World War II, the ROC had only 120 functioning churches in Russia. After the addition of new territories by the final days of the war, in 1944–45 the ROC appealed to the Soviet government to open 5,770 churches, most of which were on the newly acquired territories. Only 414 petitions were granted (less than 10 percent of the total number), but this nonetheless represented a significant increase compared to pre-war years.[19] The Easter Cross Processions from village to village were banned in 1948, and no new churches were allowed to open their doors to parishioners from November 1946 to March 1953.[20] Yet those churches that had already been opened continued to welcome their members.

In the 1950s (after 1953), the ROC still had a predominantly rural representation. Of 13,500 churches and meetinghouses, 11,500 were in the countryside.[21] Because up until the mid-1960s more than a half of all Soviet people were rural residents, and because the ROC had a much stronger presence in the countryside, the limited revival of the Orthodoxy in the 1940s and 1950s was largely a rural phenomenon. If we also factor in the post-war gender imbalance and male outer migration, this revival became a largely feminized rural phenomenon. Female workers of one collective farm in the Vologda region summed up their attitude by saying that "only God can give people something good."[22]

Women were often both the driving force in petitioning for the reopening of the church and the labor force that the church needed to reopen and survive. For example, when residents of the Fedorov village of Kirov region were granted a permission to re-open the church, women "worked day and night clearing the trash and debris from the old building, brought glass from their own house windows to put it into the church windows, and also brought paint, scrap steel, and so on," reported a representative of the Council on the Affairs of the Russian Orthodox Church in Kirov region. In a different example, women of the Kuliatino village of Salobeliakskii region "relocated on their backs 200 tons of grain, traveling a great distance, voluntarily coming to do so from different villages," and they were willing to sell their cattle to help their church financially.[23] By 1958, in the Russian Federation there were 2,403 functioning churches and 527 prayer- and meetinghouses, of which 2,222 were in the countryside.[24] Peasant women repeatedly appealed to various councils to allow their churches to reopen. The Committee on the Affairs of the ROC[25] received 1,310 written petitions and 1,700 appeals in person to open a church in 1955, and another 2,265 written and 2,299 personal appeals in 1956.[26] Even the threats to fire petitioners and their relatives from their jobs had little effect.[27]

But this religious revival and religiosity in general was experiencing a generational divide. A new generation that was raised on the Soviet socialist propaganda often criticized the older people for their religious beliefs. The most aggressive defenders of the new ideology of socialism and fighters against religion were young men. "Why do we bother with these priests?" said a worker, Marenkov. "Issue an order for them to stop all their work. Or it seems like we have two ideologies at the same time."[28] One younger woman added: "Why did they re-open the church after the war? Some religious people understood it as a return to the past."[29] And other collective farmworkers looked into the future with optimism: "It is enough to step up the propaganda among the believers and they will give up and refuse to follow any faith."[30]

The same notions and sentiments went into the decree of November 10, 1954, "About Mistakes in undertaking the scientifically atheistic propaganda among people." Its main goal was to curb the liberalization in the treatment of the ROC in the post-war years. Consequently, party representatives subjected many believers to what they labeled "a clarification of their mistakes." For example, in Archangel region in March 1955, the Head of Militia summoned "for a talk" all residents who signed a petition to reopen the church. After their undisclosed conversation, which aimed to explain the residents' mistakes and lack of proper understanding of the teachings of socialism, many residents turned in to militia their reports and letters of the following matter:

> I am explaining that we were a few women who signed something and I forgot what but then I was at home and remembered that yes, indeed, we signed something. But I take my signature back; I don't need anything like a church, they can bury me when I die anywhere, just put my body in dirt, and all my life I saw nothing good anyhow, none of the good days but only work and chores. And I have nothing else to remember. I am guilty as charged but I will get better and will never again sign anything.[31]

In another example, in Novozybkovo district of Briansk region in 1955, the director and teachers of a local school went from house to house "to talk individually" about closing the church and asked people to sign the petition for it. Some people agreed verbally but did not want their names listed. Yet nonetheless, these teachers collected 1,100 signatures in favor of closing the church. Within two days of this news reaching all villagers, they had collected 1,500 signatures against closing the church.

But the authorities did not totally eliminate all religious life in the country-side, even by making the believers pay fines, threatening them with losing their jobs, or confiscating their gardening plots, as well as by waging propaganda campaigns against teachings of the Church. Moreover, the Church performed a social function by providing public space for the whole community to meet and socialize. Local clubs and the so-called houses of culture that had after-hours hobby clubs or movie showings, were supposed to be an alternative space that would challenge and replace the social function performed by the church. The government assumed correctly that, in the countryside, the churches satisfied not just a spiritual but a social need as well. Yet it misjudged how the social function was also multi-dimensional, and churches and clubs could co-exist without necessarily competing with each other. And this is exactly what happened in the countryside. For example, as of 1954, 23 villages of the Ulianosk region had churches and mosques as well as clubs and houses of culture. On all occasions of religious festivity, the doors of the clubs were open to entertain villagers with concerts and movies. Yet religious holidays and festivals attracted hundreds, and at times thousands, of villagers, who went to clubs afterwards or on alternative days.[32] For these people, the two places, the church and the club, played such drastically different roles that both became part of everyday life without competing with each other.

At times, political enthusiasts went from village to village to lecture about the harmful ways of religion and the advantages of the Soviet order. But in the 1950s, the number of such enthusiasts was insignificant, and they were often met with hostility when they started speaking about the great achieve-ments of the Soviet Union. One villager from a region where only 10 percent of all rural houses had electricity and only a third of all rural houses had radio, spoke up: "What are you talking about, Comrade lecturer? My father suffered under tsarist regime, and I suffer under the Soviet power."[33]

The anti-religious campaign was reinforced legislatively. In 1961, a Decree of the Soviet of Ministers "On the tightening of control over the affairs of the ROC," as well as a Decree "On monasteries in the USSR" and "Taxation of Religious Organizations and Monasteries' Income," aimed to break the limited revival of the ROC. Moreover, all legal provisions and rights that were extended to the Church in the 1940s were officially annulled. Clergy were removed from handling the administrative and fiscal affairs of the ROC. The ROC was banned from any charitable activities and from engaging children under the age of 18. Parents in their turn were "advised against bringing children with them to the church."[34]

For the government, it also became convenient to blame the Church for mass abstentions from work during religious holidays, which were also often accompanied by heavy drinking. The Church was blamed with promoting alcoholism in the countryside. But local residents were not easily swayed, but instead believed that "indeed, the cases of alcoholism are common in the collective farms but they have nothing to do with religion. If people drink, they drink without crossing themselves and while forgetting God. Drunken holidays do not exemplify people's attitude to religion."[35] Drinking during holidays and various festivities was in fact common, but that applied to all holidays, not just religious ones. May 1 (International Workers' Day) and November 7 (Anniversary of the October Revolution) were equally—or arguably even more so—an occasion to drink as any religious holiday.

The government, party and collective farm officials also feared that any increase in income among the population would result in greater expenditure on candles and donations to the Church.[36] It almost became routinely expected of party members to criticize priests and "expose" the abuses of these "legalized lazy bums." The priests were accused of "going around to appropriate huge sums [of money]" for doing nothing. Yet local residents knew their pastors well and rose to their defense, at the very least by implementing none of the measures against local priests proposed by local party cell leaders.[37] Accusations of enriching themselves unfairly were also levied on the lay members of the Church, mostly elderly women who cleaned and maintained church buildings and often managed miniature Church stores that sold icons and candles. Such women were often accused of being "profit-thirsty" and undermining social equality in the socialist state. However, such accusations disregarded the fact that many lay members donated most of their income to the Church.[38]

But once again, some women took it upon themselves to oppose the closing of the church or to demand its re-establishment. In one of numerous similar cases, the Soviet of Deputies in a village in Tambov region decided to take down the church and use scrap materials for repairing a local school building. Officially, the church was listed as "in poor, dilapidated shape" and thus only suitable for demolition. But rural residents decided to have a round-the-clock watch to protect the structure and not let anyone in. A total of 40 women moved into the structure to stay there until the decision to demolish it was reversed. About 200 other women brought money for repairs, as well as personal items to use in the church. In the end the Soviet reversed its decision, and women demanded and received a right to have regular services in the church.[39]

In the countryside, the ROC continued to perform many of the traditional rituals, such as baptisms, weddings, and other special services. Numerous checks and inspections conducted by the government and party authorities revealed that baptisms were still common in the countryside. One such inspection demonstrated that in Kalinin region every third child born in 1959 was baptized.[40] As a consequence, the Church revenues in that year reached 10 million rubles, or three times the state income from taxing the residents of this region.[41] In Sverdlovsk region, 2,000 children were baptized between January and March 1956, and 2,400 more between January and March 1957. In Karpovskaia church of the Gorkii region, at times the number of baptisms in a single day reached 145.[42] Also recorded were cases of baptizing children who were born into a family where both parents were party members.[43] Although less common than baptisms, church weddings were also present, including a wedding between two members of the Youth Communist League (Komsomol).[44] In Sverdlovsk, again for January–March 1956, 457 couples were married in a church, and 525 during the same months in 1957. In Gokii region, up to 150 couples a day got married in the Karpovskaia church in 1957.[45]

Peasants' religious faith was so significant in the countryside that some people even relied on collective prayers to address everyday problems such as droughts or poor harvests. In the state farm "Selikhovo" of Torzhok district, the peasants asked priests to do a cross procession to plea to God for rain at the time of drought. The priests were more than willing to grant the wish.[46] Local party members explained this by saying that "the Church used people's need to its own end, and these old [clergy] were patient and polite." At the same time, these same party members also had to admit that many members of the clergy were in fact "young priests" (and not "old" members) and that they were "well educated, polite, and demonstrated love for their work."[47] The parishioners responded to the clergy's care with respect and willingness to listen to the clergy's advice. For peasants this was a rare opportunity to speak their mind and complain about the harshness of their rural lives. They craved solace and spiritual peace. Collective farm workers in Moscow region complained that "in our country much is written and said about the freedom of religion but it does not exist in practice. If we do not preserve our church, we will stop working for a collective farm."[48] But once again, local party representatives blamed insufficient propaganda and poor entertainment offered by local clubs for the survival of these religious sentiments.[49]

This confrontation between the Soviet authorities and the ROC and its believers continued throughout the years on all levels. Often, local authorities felt that they had full and complete rights to control all the affairs of the Church and of its individual followers.[50] And in many cases, only the Council for the Affairs of the ROC could intervene. Maria Dmitrievna Eremeieva requested the presence of a priest to perform the Sacrament of Holy Unction (*soborovanie*). But the head of the local village party cell (*sel'sovet*) wrote the following to the sick elderly woman:

> Comrade Eremeieva, the Soviet orders you, once the priest arrives, to show up together to the soviet. If you do not show up, you will be fined 500 rubles; and the priest has to show his documents that he has been allowed to enter our village.[51]

After the appeal to the Council for the Affairs of the ROC, the Council intervened in this case and found the directive "unethical."

Most commonly, local authorities ordered that gas and water for the use of the church be turned off "in order to limit the functioning of the church," or did not allow any repair and maintenance work in buildings or on the grounds. It was proposed in Kalinin region that the church building was blown up before peasants had a chance to appeal for its reopening. In Tula region, collective farmworkers were required to attend regular meetings in order to discuss the need to close churches.[52] In Birskii district the Head of the district party cell ordered that the church be closed. Upon receiving the order, the head of the collective farm, director of the village school, director of the village club, and the secretary of the party youth organization cell in the village organized a meeting with collective farm workers to talk about the order and explain its benefits. Peasants were outraged, and the most ardent believers, mostly elderly women, appealed to the Council to complain about the decision and collected signatures to protest the closing of the church.[53] Similar cases were common everywhere.[54]

The "church reform," initiated by Khrushchev and lasting into the 1980s, aimed to strip the Church of its legal status and to limit the rights of the clergy to control fiscal and administrative affairs. In addition, it also emphasized the need to prevent children from being exposed to the teachings of the Church. In Rostov region this resolution was implemented the following way: the local resolution stipulated that soviets were "to take measures to assure that children and youth under the age of 18 are not allowed to go to the church and prayer-houses, and neither are allowed to have any services and ceremonies performed for them."[55] As a consequence, on April 4, 1965, members of one Rostov church started to scream and protest in public when they were prevented by a local militia from entering the church with their children . The same protests were reported for churches all over Rostov region. The regional party cell demanded that the Church provided it with a special form for each and every case of children's baptism.[56]

When it came to defending themselves, some believers complained openly about their rights being violated. The Council of Ministers received 57 written complaints in the first four months of 1965, and saw seven delegations that appeared to complain about abuses in their villages. One complaint said:

> Why do they [local authorities] treat us this way? Please, help us to find justice. We do not want anything unlawful and we do not do anything unlawful. We are neither fascists nor enemies of the people; we are faithful believers who pray for the wellbeing of our children, our beloved motherland, our government, and peace in the entire world. We are old women; we do not have much life ahead of us . . . Please, open our cathedral.

Such petitions were signed by hundreds of people.[57]

Ironically, the closing of selected churches and attacks on religion had the opposite effect to what the government intended. The remaining churches performed more services than usual and had higher revenues than ever before, and people seemed to be drawn to the Church regardless of the official viewpoint. Thus, even though the number of churches in Rostov region went down four times from 1960 to 1964, in 1960 only one in five children born were baptized, whereas one in three children were baptized in 1964. The income rose steadily as well; in 1962 the ROC revenues in Rostov were 1.42 million rubles in 1962 but rose to 1.53 million rubles in 1964.[58]

The same trend existed across the country. In one village in Tambov region the village council decided to close down the church in 1965, justifying their decision by citing the poor and unsafe condition of the building. But the collective farmworkers went on strike over the decision and refused to return to work for several months, while also keeping a round-the-clock watch on the church building. As a consequence, the decision was reversed. In three villages of Kalinin region believers were threatened with losing their jobs and their gardening plots if they did not quit the Church, but they did not. In Yaroslavl region two young parents who came from a local village to study at a teacher-training college baptized their child. They were expelled from their school, but demanded justice and were reinstated once the city council intervened.[59] In Penza region, a review of baptismal records revealed that in the majority of recent services, parents who brought their children to Church were active members of the local youth organization and some were communist party members.[60] Even though the mistreatment of believers and abuses of the Church continued, Chair of the Council for the Affairs of the ROC had to explain them away in a top-secret document written on January 27, 1966, by saying that "these misdeeds most commonly result from a poor knowledge of the Soviet laws by local party officials and from an inefficient supervision of proper implementation of these laws by respective government organs."[61]

The tense relationship between the state and the Church, and a slow but systematic curbing of the ROC's rights and position, continued well into the 1980s. The overall number of churches in 1986 was only half of those that functioned in the early 1950s. Equally, almost all aspects of the religious life were overseen and controlled by the Council for the Affairs of the ROC, even if it was renamed as the Council for Religious Affairs, in order to acknowledge and supervise other religious institutions in the Soviet Union.[62] Looking back at the 1960s, 1970s and early 1980s, one of the ROC leaders called this Council "an ulcer on the body of the Church and the state."[63]

But reforms initiated by Mikhail Gorbachev did not leave the ROC off the agenda. Openness and restructuring (*glasnost* and *perestroika*) eventually reached the Church affairs as well. The symbolic turning point in the relationship between the Church and the state was the celebration of the millennium of Orthodoxy in Russia in 1988. In the same year, the ROC was granted a right to proselytize, teach, publish religious literature, perform charitable actions, and conduct services freely. The subsequent transformation of life, along with other freedoms, brought a greater freedom of religion. According to the All-Russian Center for the Study of Public Opinion, in 1993 over a half the population of Russia considered themselves religious, and over 90 percent of these belonged to the ROC. By 1989, 60 percent of all residents of the Russian Federation were baptized, and this percentage increased to

75 percent by the end of 1992. But many believers still did not attend the church regularly; in the countryside, the most common explanation was the distance one had to travel to reach a church. Among the believers, the two predominant groups were elderly women and young men and women, usually in their late teens and early twenties. Among the elderly, most people had an incomplete secondary education, but the younger generations of believers were well educated.[64] Religion was listed as the single most important aspect of life by 3 percent of women in cities and 6 percent of women in the countryside.[65]

In the twenty-first century, the age difference among believers has become less pronounced, and all generations agree that the main goal of any religion is to teach its followers moral values and social responsibility.[66] In addition, 32 percent of women and 26 percent of men believe that there is no need for the separation of the Church and state, and that the Russian Orthodoxy should have the status of Russia's official religion.[67] The active role that the Church plays in local communities, and its charitable actions, were noted as its main strongholds. Although not all people who "believed in Orthodoxy" listed themselves as Orthodox, 65 percent of women and 40 percent of men in Russia identified themselves as followers of the ROC.[68] Christmas was also cited as one of the most important holidays, following New Year, Victory Day and the March 8 (Women's Day).[69] This common acceptance of the ROC also changed the prevailing notions about ardent believers. During the Soviet days, the common point of view was that only elderly, socially displaced or politically and intellectually incapable people turned to religion for salvation. But this notion was replaced with new understanding that the Church is a safe haven for those who, in addition to seeking faith, oppose violence and want a healthier society devoid of ruinous addictions and obsessive materialism.[70]

Some younger people also want to escape the challenges—or evils—of present-day society by taking vows and joining monasteries. As of 2003, 500 monasteries housed thousands of monks and nuns, and the number of nuns' convents grew rapidly in the following years as well.[71] Also in 2003, the total number of religious organizations (including ROC and others) was 11,299, and this number increases every year.[72] As one woman aptly phrased it, "we waited for this our entire lives."[73]

16 Triple-burden lifestyle

A woman's work is never done.

According to a long-standing tradition, it was a woman's responsibility to take care of family's private garden plot and any domestic animals a peasant family might have. This tradition continued well into the Soviet era as well. When the Soviet government announced the establishment of collective farming, more men than women became involved in the process. In the 1930s, two-thirds of all labor days, a common system of payment that calculated the number of days each person worked on the collective farm, were completed by men.[1] Women's work in the collective farm was often seasonal, and women still spent more time performing domestic chores and working their private plots. Thus, in the 1930s, up to 80 percent of all hours worked by women were not spent at their workplace. The work on the private lot also supported the family by providing it with extra income and a bulk of vegetables, eggs, dairy products, and even meats for its consumption.[2]

In the aftermath of World War II, private lots became the only means of survival. Collective and state farms worked almost exclusively to support the state and to meet its production quotas, while most produce for personal consumption was grown by peasants during their after-work hours. In 1946–47 rural residents even resorted to growing their own grain, which took up as much as 15 percent of all land in personal use, although the potato was still the crop of choice. The war tax on private plots was liquidated after 1947, but taxation in kind or cash for peasants with private plots continued. The Department of Correspondence and Public Relations of the Supreme Soviet of the USSR had barely enough resources to register all letters, complaints and appeals from rural residents who saw extra taxation as unfair.[3] In one such letter, written in 1952, O.P. Zhideleva complained:

> I work at the animal farm in the collective farm. I work all year long and without any weekends. My daughter also worked all of last year, but we cannot make enough to feed ourselves. I also have two children who are not yet in school. How are they to be taught? And how will I feed them?

I don't know. I have nothing. I earned 500 labor days last year but was given only 140 kilos [of flour]. We do not get money for labor days. Whatever we earn by selling milk [from our own cows] goes to buy more flour and pay taxes. Other two children have to drop out of school because there is no food, no clothing, and no shoes. What do we do, Comrade Stalin? We work all year long and do not earn even for bread! I get so upset about this: we grow grain [for the collective farm] yet we have no bread to eat for ourselves.[4]

In 1946–50, when the Soviet Union threw all its resources to rebuilding after the war, very few collective farms paid adequately for their workers' labor, even though much of that labor was female and much of it had to be manual, due to the lack of machinery. The best of the best collective farms were used in massive propaganda campaigns to demonstrate that such exemplarily farms as "Krasnyi Putilovets" in Kalinin region paid up to 3–4 kilograms of grain for each labor day.[5] Movies such as "Kuban Cossacks" were shot with an unrealistic array of foods at each meal in order to demonstrate the abundance of grain and other foods available to peasants. But in reality, almost a half of all collective farms gave less than 1 kilo of grain per labor day, and 40 percent of farms paid less than 60 kopeks for a labor day.[6] In some farms, workers were not compensated for their labor days, and had to go without any pay at all for their work.[7]

Equally problematic was finding hay for privately owned cows. Cows were the life support of most families, partly because the milk was used for dairy products and partly because of the potential for producing calves and subsequently making profits on their sale. Legally, peasants could get up to 10 percent of their pay in hay, but throughout the 1950s refusals to give peasants any hay at all were common. Such refusals were typically justified by the fact that there was not enough hay to feed even state-owned cattle. As a predictable consequence, the peasants' last resort was to steal hay or to sell their livestock. Peasants explained that they had "objective reasons" to sell their most prized possession.[8] The long and painful process of post-war reconstruction and the systematic refusal to give peasants any hay resulted in a significant reduction in the number of cows owned by individual rural dwellers. In some regions, thousands of households ended up without cows, and by the late 1950s most households had no livestock at all.[9]

The famine of 1946–47 demoralized the country's citizens, both rural and urban. But in large cities the rationing of food made basic provisioning—ironically—even easier than in the countryside. The information about living conditions in rural parts of the country that the mass media presented in cities, especially in Moscow, poorly reflected the reality of rural life. When many peasants were starving, one correspondent based in Moscow wrote that "though the Soviet agriculture indeed suffered in the war and as a result of last year's drought, it is doing just fine in this tough time now thanks to our superior system of agriculture and its effectiveness."[10]

But, in reality, most rural people starved and were forced by an inadequate system of payment for labor days to spend progressively more time on their private plots to assure their mere survival. The state continued to extract most of the agricultural produce while paying little or nothing at all for it. Some people, and especially the younger generation, opposed the economic policies of the state by running away and migrating to urban centers. Organized mass migration became the norm of the day for those who saw it as the only form of passive resistance available to rural people.[11] But others invested the bulk of their time and effort into their own piece of land to grow at least enough for their meager rations. Thus, in a collective farm in Vologda region in 1947, 50 people worked on the farm "regularly," but only 12 of them worked daily. In addition, 53 workers were supposed to work on the collective farm but worked on the railroad or elsewhere, while 11 more worked in a nearby local city center "without explaining anything." Those who worked on the collective farm earned on average 163 labor days annually. This number is not a fair representation of what really happened on the collective farms. The director and other state officials of the farm earned 577 labor days a year, even though they never actually worked in the field. Vets and those who worked in the animal farm sector earned 378 days, but field workers who did all farming from plowing the land to harvesting, earned a meager 88 days. There were also many workers, primarily women, who had no gainful employment and spent most of their time working on their own land allotments.

Poor payment for labor days created a vicious cycle: the farm paid little or nothing for labor days, thus peasants had no will to work for the collective farm, thus more effort went to the private lot, thus less work was done on the farm, thus the farm would pay even less or nothing at all, and the cycle started anew. This tendency was also confirmed by official taxpayers' records. The income of each village was inversely proportionate to the number of people who worked on the collective farm. The highest income was reported for those villages where all able-bodied people worked but the income from the collective farm was minimal (e.g. 259 rubles of the total amount of 14,000 rubles).[12]

As a consequence, a new policy of limiting private garden plots was introduced and enforced in the late 1950s and early 1960s. Limits were set for the size of private allotments and the number of cattle heads in private ownership. The number of cows, sheep and goats decreased, but work on private plots remained crucial and central to rural life, especially for women. In 1959, 10 million people worked exclusively on their private plots and had no other employment. Over 70 percent of all rural residents worked on their private allotments after work, and of all those who worked the private land 90 percent were women.[13] This was a female task that added yet another aspect to an already double-burden life of a woman who had to carry on both paid work and nearly all domestic chores. For rural women, the life was a triple burden, with cattle and a private allotment adding another dimension to child caring, cooking and cleaning, and gainful employment. And these women were remarkably productive, as they used 4 percent of all arable land in the Soviet Union and yet supplied the state in 1959 with 65 percent of all potatoes, 40 percent of meat, 50 percent of milk, and 80 percent of eggs, paid as taxes in kind.[14]

The campaign to limit private household use of land had dramatic consequences. In 1959–64 the land usage declined from 450,000 to 380,000 hectares, and the contributions to the state declined proportionally. In an attempt to bring up the numbers, all restrictions were lifted in 1964. By 1970, because of the persistent encouragement from the government, the acreage and production on gardening plots returned to its pre-restriction levels. In the 1970s, 90 percent of rural families in Siberia had personal plots, and two-thirds had cattle, livestock or some domesticated animals. Families that did not have any land for personal use were considered "out of norm" and unusual, and work on the plot became almost a tradition in a Siberian village. The plot provided more than just supplementary food for the family; it often provided families with their main ration. It also was a crucial source of food for family members who lived in larger urban centers. Sometimes for the sake of saving money, but more commonly out a dire need, these urban dwellers got much of their caloric intake from produce sent from home. The allotment was also a source of additional income, and rural families often pointed out that it allowed them to do well financially despite abysmally low wages paid by collective farms. But the benefits of having the land did not ease the challenges of working that extra piece of land, often after the working day and on the weekends and holidays. In the 1970s, various social studies pointed to this discrepancy: villagers wanted to *ease* their lives and have more money by working *hard* on their personal plots.[15] And indeed, enough families wanted to work on their personal plots to make it into a common everyday reality of the Soviet countryside. Although women contributed the majority of the labor, all family members were tied to the land and worked extra to ensure the family's prosperity.

The gendered labor division that existed in the countryside was clearly present in this case as well. Men assumed responsibilities that were considered traditionally male and labor intensive. Thus, most commonly men were responsible for building, maintaining and repairing barns, wells, and other structures. They also partook in turning the land after winter, digging potatoes, and at times, watering (because one had to haul water from afar or lift buckets of water from a well). Women typically did the rest of the work in the garden, and later processed whatever they had grown to ensure long-term storage. Jams, preserves, pickled and dried vegetables and fruits were all common choices. Taking care of any livestock was the sole responsibility of a woman, which added significantly to her workload. Of all labor, women in general invested 35 percent of it into the personal land and farm, whereas men only invested 9 percent.[16] On a per-hour annual basis, in 1965 women invested 589 hours of labor into their personal farming and husbandry, whereas men spent 112 hours. In the 1980s the gap had shrunk to 549 hours for women and 219 hours for men, mostly because livestock became less common among rural households. Even retired men and women shared the workload unequally, with 921 hours worked by women and 305 by men.[17]

Policies of the state remained contradictory for the last 20 years of the Soviet regime. On one hand, as mentioned above, the state encouraged the revival of personal farming, starting in the mid-1960s. But on the other hand, this encouragement had immediate implications, and more and more resources were invested into attracting rural people to work on the collective farms. In 1965, collective farmworkers were finally granted the right to state pensions in retirement, and a right to monetary wages when they worked on the collective farm. In the 1970s, 34 million households still had land and livestock for their personal use, of which 13 million were collective farmworkers and 10 million were state farmworkers.[18] But the lure of guaranteed wages and pensions in the collective farms, along with steady outer migration of the younger generations, attracted to the promises of urban life, ensured a slow but steady decline in the numbers of households with private land.

The demise of the Soviet Union radically undermined the entire structure of rural life. Within a matter of 2 years the system collapsed, and along with it disappeared such features as regulated prices, guaranteed employment, social equality (even if at low levels), and secure old age. The new model of farming, based on a model of individual farms similar to what exists in most developed countries such as the US and countries of Western Europe, was not easily received by rural residents. In the early 1990s, more than half of all women in the countryside said that they preferred to keep collective farms and guaranteed wages, even if these wages were to never go up or improve. Most women with grown-up children reported that they did not want their children to become private commercial farmers. For these Soviet people, private garden plots were a way to *supplement* their incomes, not a way to make one.[19] Rural women became the conservative force that opposed the introduction of new farming styles and models of organization.

The insecurities of the new age made change all the more daunting. Many peasants believed that private land ownership for commercial farming was "wrong"; that "one cannot survive alone in the countryside. We are all farmers in what we do but we cannot survive without a collective farm." Another voice echoed the same sentiment that "even if there were a wealthy commercial farmer nearby who paid well for work and people went to work there, I would not go."[20] Moreover, women were often concerned that new farming organization required more work on their part, because traditionally women, not men, invested the most time in providing for the family. They were already overwhelmed with responsibilities of caring for children and often elderly parents, in addition to carrying out gainful employment and working their plots. Hence these women were cautious about the prospects of adding more work to their already burdensome workload.[21]

As a consequence, two-thirds of all private commercial farms were established by urban dwellers who wanted to experiment with new entrepreneurial opportunities.[22] And even of those farmers who were born and raised in the countryside, the majority of new proprietors were people who resented a collective work approach and were not team players in any setting. They carved out not the opportunities for personal advancement but a possibility to make a living without a boss watching over their shoulder.[23] Yet these former city residents did not plan on spending their lives in the village. They wanted to see what this new commercial farming was all about but the final pressure to start a new business typically came after they lost their jobs and consequently felt significantly dissatisfied with their lives.

The early commercial farms were also far from perfect models of agricultural production. The Malikhonov family decided to start their own business by moving from a local urban center to a village to raise calves. At first, the family had 20 calves to take care of, but only two months later they added another 388 to that number! The farm seemed to take off. But the mother of the family complained of the heavy labor, saying "at first, we only had rakes and shovels as our tools."[24] Everything had to be done manually, including cleaning the manure and bringing up to 200 buckets of water from a well to the cattle ward. Elena Malikhonova added that they were treated poorly by their neighbors and no one wanted to follow their example. "We are like a thorn in their eye," said Elena. "People fear that we will earn too much, get too rich. But we will earn, not steal! Yet no one wants to do what we do."[25]

Of those who risked starting a commercial farm, not all stories had a happy end. Women were disappointed with the results of their initiatives, and 40 percent of women in private farming reported that their expectations were not met at all. They emphasized the difficulty of manual labor as well as the fact that "it is tough to survive in a world of men's business."[26]

Some opposition to such initiatives even came from within the families. For example, Raisa Shaikhieva decided to open a farm and move from a city to a village. Her husband and children opposed her decision to leave the comforts of the city life. Even though such comforts were meager, they wanted an easier life than that of a farmer. But Raisa decided to persevere. She had been fired from two jobs, one after another, and this was enough to make her seek other options. She moved to a village by herself and started an animal farm with 60 calves. She worked alone, and did all the labor manually. She worked the land to care for the pasture and to grow some staple foods such as potatoes, and she had to bring water from a well far away to water the land, clean and wash her calves, and give water to her cattle to drink. On some days she passed out from exhaustion. Raisa could not leave her farm for travel or vacations for 5 years, not only because she was busy but also because she had only a single pair of rubber shoes. And, to add to this workload, she had one of her children with her who was only 1 year old when Raisa started the farm. The farm became profitable 5 years later, and soon thereafter Raisa's husband and older children decided to move to live with her. But the amount of labor proved too challenging for many.[27]

Such initiatives did not effect or undermine popular reliance on relatively small garden plots. Even in 1990, the last year of the Soviet life, small allotments continued to provide rural residents with two-thirds of all the potatoes they consumed and at least a quarter of all milk.[28] The economic instability that accompanied the post-Soviet economic depression made individual allotments even more important for food consumption and overall agricultural output. In 1993, there were about 50 million families who worked their plots to supplement their incomes, of which only 13 million parcels belonged to urban dwellers. Their contribution to overall agricultural production rose from 22.5 percent in 1990 to 36 percent in 1993.[29]

In 2003, all private farming was divided into three general categories:

- Small private garden plots; 16 million families had on average 0.4 hectares and another 19 million families had on average 0.09 hectares in their private use. They collectively grew, mostly for personal consumption, 53.8 percent of all agricultural produce grown in Russia!
- Agricultural organizations on a collective basis (cooperatives, limited responsibility organizations, and others), with 171 workers per organization on average. They produced 42.2 percent of all agricultural output of the country.
- Private commercial farms, numbering at 265,500, which produced 4 percent of all agricultural output.[30]

Most people who work private plots in the countryside are of the retirement age, and demographically, women outnumber men in this age group by a factor of two (they outnumber men in all age groups, standing at 18.4 million men to 20.3 million women in rural Russia).[31] Russian villages experienced a massive aging process, when most young men and women went to study in local towns and never returned.[32] It also experienced an even further feminization of rural labor. The state policies of recent years were to encourage personal farming, including all possible forms of agricultural production and organization.[33] In addition to legislative changes to foster private initiative in the agricultural sector,[34] the government also pays subsidies to the poorest rural residents in the hope of preventing them from migrating to cities. These subsidies reached 2 billion rubles in 2003.[35]

Women also explained their greater involvement in working their private lots (compared to men) by citing their "traditional and age-old habit of working around the clock." Antonina Mikhailovna Taryshkina from Saratov region commented:

> I am 83 years old, my days are gone but I still keep everything in order ... I work all day and then see, oh my, it's getting late! Why is the day so short? My mother used to say that only a lazy person gets ill or feels any ailments, but if you keep busy, you might still get ill but you do not

notice it. And when I get up in the morning, I feel my age; everything hurts! But then I get to work, and it seems to get better.[36]

The overwhelming workload of rural women remains undeniable.[37] But women themselves explain their position in ways similar to how one woman summed it up: "I'll tell you: we, women of Russia, spoil our men. We do everything by ourselves!"[38]

17 Household chores

The "everything" that women did for themselves undeniably included household chores. The notion of the double burden that most women carried, which combined paid work with being solely responsible for domestic chores and child rearing, has been well recognized and studied by historians. For rural women the burden was triple, adding work on a garden plot to other responsibilities. But much of the work done by women was never fully appreciated or even recognized. While an average income and wages for women were comparable to average wages for men, everything in the domestic sphere was considered to be "a woman's work" regardless of whether this woman had a full-time job or not. Even women in highly-paid professions with advanced degrees spent on average twice as much time doing their household duties as their husbands. In addition, the amount of help offered by a man was, and continues to be, considered a personal choice—or in other words, if a woman got no help at all, it was her own problem.[1] Moreover, social studies have demonstrated that married men intentionally used women's unaccounted work around the house to relegate all domestic responsibilities to their wives for the sake of having greater control of their own time and better careers.[2]

Not all women accepted their position, yet most protests were limited to occasionally asking a man to give a hand around the house. But years of gender equality propaganda, and later information about gender equality in other developed nations, created a generational rift in attitudes to domestic chores. While women continued to do the majority of work around the house, they nonetheless became more critical of this situation than their mothers and grandmothers.[3]

Women of the older generations typically assigned men a status higher than for themselves, simply because, as they explained, "men were men." This attitude of revering males was applicable to most aspects of life, from work to personal interactions. One elderly woman summed it up by saying that "a man is better than a woman; a man has brains because he is a man."[4] Men had an equally high opinion of their status. One man, born 1915, was asked to explain why he believed himself to be the last resident of his village in Vologda region when there were three other women there. The man was at

Figures 17.1 and 17.2
Burekhina and her
family at home after
a day of work at a
collective farm "Zaria,"
in 1947, Russia,
Sverdlovsk region.

Source: © Courtesy of the
Russian State
Documentary Film and
Photo Archive,
Krasnogorsk, Russia

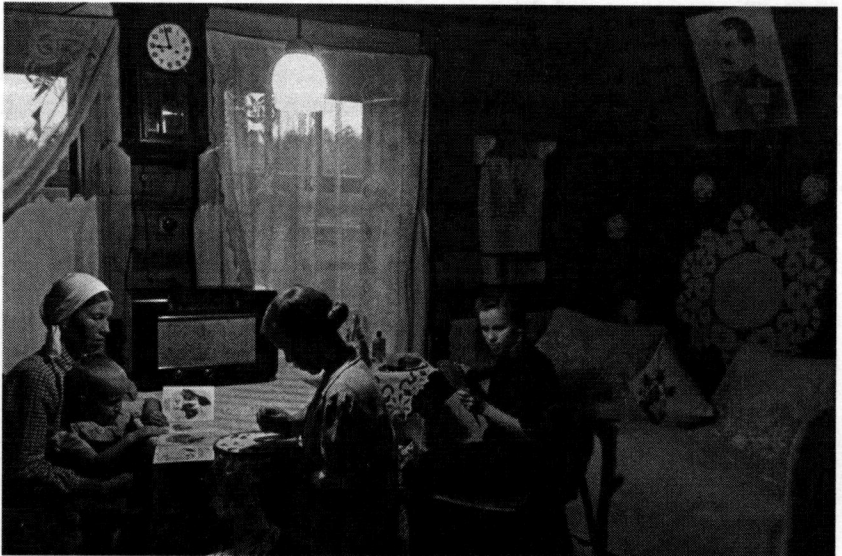

first puzzled by the question and then explained that "women do not count, of course."[5]

The bulk of all domestic chores was performed by women, especially such chores as cooking, doing dishes, cleaning the house, doing laundry, and taking care of children. One of the most labor-intensive chores was washing the floors, always by hand (with no mops), and in most rural houses, if the floors were not painted, the floors had to be scrubbed with a utensil that resembled a knife. Dirt floors, which were common in villages of Tambov region even in the 1960s, were not much easier to take care of, and all floors were covered with rugs that had to be regularly washed in a nearby brook or river.

Laundry was done using a washboard in a washtub. After washing, everything had to be ironed. For clothing, the irons were cast-iron with a space to put hot coal in to warm it up. Needless to say, such old-fashioned irons were heavy and hot, and there was always the risk of getting some coal on the clothing. There were also irons that could be warmed up on a stove, but they cooled off fast and took a long time to warm up again. Thus, such ironing was usually more time consuming than by using a coal iron. When electric irons were introduced, they immediately became a much sought-after item. But they remained a rarity even in the 1960s, and most rural residents continued to iron the old-fashioned way.

To do all the chores, women typically spent three to five hours a day doing housework, or even longer if they had to make their own bread (compared to half an hour spent by men doing household chores).[6] Some men were sympathetic to the hardships of their wives, yet such sympathy never translated into an effort to help around the house. Men often justified their inaction by saying that this was the way things were done in their parents' home and thus should continue to be so. Yet they forgot to mention that, unlike in the case of pre-Soviet Russia, most women in the Soviet days added a full-time job on a collective farm to their domestic chores.[7] Gender imbalance in the post-war years, when most losses in the war were male, made women more inclined to tolerate men's "weakness." There was a common notion that it was better to have any man than none at all, and women "who were taught by the war to love life and see its beauty" were willing to accept the bulk of the work for the sake of having a family.[8]

The devastation of the war years was not easy to overcome. But by the late 1950s and early 1960s, the countryside experienced a construction boom. Most dugouts were replaced with permanent houses, and most houses boasted electricity and radios. When there was any choice, rural residents preferred to have individual houses over flats in apartment buildings, mostly because such houses allowed them to keep and work on gardening plots.[9] As a result, such individual houses constituted 80 percent of all new housing in the countryside.[10] But separately houses were not meant to have modern conveniences. Even apartment buildings often lacked central heating, running water, and any sewage.[11] By the late 1980s, only half of all houses in the countryside had running water, and only a third had central heating and sewage. Only one-sixth of all houses had hot water, although it was accessible somewhere in the proximity in 80 percent of all cases. But for houses constructed in the interwar decades, most modern conveniences remained an unheard-of luxury, even in the 1990s.[12]

The lack of running water and central heating shaped the workload of all families, especially women. For example, even in the 1990s in Novgorod region an average rural family spent 12–14 hours a week sourcing water for its needs, and invested two months of overtime labor into preparing and storing wood for heating. Not even counting laundry, a family needed 10–12 buckets of water daily, all of which had to be hauled from the well, which at times was located several houses away.[14] Some villages were even less fortunate. Residents of Vasilievskaia village of Orlov region had to walk 2 kilometers to get water, and they resorted to melting snow in the wintertime.[15] In Yakimovo village of Kaluga region, women had to climb a cliff to get water and they had to use ropes to pull up buckets full of water. It is not an exaggeration to say that they risked their health, or even their lives, in this endeavor.[16] "There are no conveniences in an individual house," wrote one woman to a woman's journal,

> Everything has to be done with one's own hands. We envy urban people; they go to their jobs during the day but then they come back home, and

Figure 17.3 Young women from a local collective farm in a village of Nastas'ino at the well in 1948, Russia, Moscow region.

Source: Photo taken by N. Biriukov.© Courtesy of the Russian State Documentary Film and Photo Archive, Krasnogorsk, Russia

it is bright and warm there, and there is running cold and hot water, and there is gas. But here in an individual house, you come back after work and it is just the beginning; you need to sweep the chimney, bring coal or wood in, warm up the house and the stove for cooking, start making dinner, and before you know it, you already work more than a second shift.[17]

Statistics confirm this sentiment. On average, a woman who lived in a city spent 1 hour 13 minutes cooking dinner, where as a rural woman spent more than 1 hour 30 minutes. During a work week, a woman in the countryside had at most 1 hour 57 minutes for activities other than work and house chores, and only 4 hours 54 minutes on weekends. By contrast, a woman in a city had 3 hours 13 minutes a day during work week and 6 hours 18 minutes a day on weekends. This difference was enough to affect lifestyles of women—urban women had time for movies, reading, and socializing, whereas such activities were largely out of reach for peasant women who simply did not have time for them.[18]

Radio was one of the few options available to peasants for recreation. It was entertainment, and it was often the only source of information about the outside world beyond the boundaries of a village. But even radio was not universally available, despite its relatively rapid spread.[19] In 1970, one-sixth of all villages still had no radio in the non-Black Soil regions of the Soviet Union.[20] But the presence of radio did not guarantee that any program was actually aired. More often than not, radio transmission was of a poor quality, with abundant background noises, and programs were not aired around the clock. Villagers complained in a letter to a magazine that their "radio was either silent or makes all sorts of noises, like a woodpecker on a tree. And

Table 17.1 Time allocation in 1934 and 1973–74, gender-specific, in hours per person on an annual basis

Type of activity	Investment of time			
	Men		Women	
	1934	1973–74	1934	1973–74
Paid work (i.e. on a collective farm)	3,168.0	2,971.7	1,152.0	2,185.3
Employment-related activities	360.0	307.2	288.0	310.5
Labor on a private gardening plot	432.0	732.5	3,024.0	1,872.4
Total	3,960.0	4,011.4	4,464.0	4,368.2
Time spent on physiological needs	3,600.0	3,516.6	3,456.0	3,396.3
Free time	1,044.0	1,102.8	684.0	867.8
Other and unaccounted time	36.0	9.2	36.0	7.7
Total	8,640	8,640	8,640	8,640

Source: *Trud v SSSR (1934)* (Moscow, 1935), pp. 378–9; *Biudzhet vremeni sel'skogo naseleniia* (Moscow (1979), p. 106

not long ago it was totally turned off because we do not have enough poles
... Now we have nothing for our spare time."[21]

Lack of electricity was an even bigger problem, a real "big tragedy" of the
countryside.[22] In the 1950s, only one-fifth of all houses had electricity in the
Pskov and Penza regions.[23] In Pskov region, in the mid-1960s, two-thirds of
all houses still lacked electricity and used oil lamps for lighting, often without
protective glass and therefore potentially fire hazardous.[24] In 1970, 10 percent
of all villages in the non-Black Soil regions still had no electricity.[25] On
occasion, some villagers got so used to their lack of electricity that they no
longer wanted it. One elderly woman refused to have electric lights in her
house in the 1990s, especially around her icons, and always asked her relatives
in a nearby town to bring her protective glass for her oil lamp.[26]

Household items had to be purchased somewhere, and stores were of crucial
importance for acquiring everything from salt and bread to clothing and
furniture. In most large villages, there were stores that sold everything from
food to items of personal hygiene to household items and wedding gowns.
Stores also became places of socializing where locals exchanged news and
rumors about their neighbors. But even this basic feature was often denied to
rural residents, and many smaller villages lacked even rudimentary stores.
Peasants were forced to walk or travel to stores miles away from their homes,
and at times were not even able to do a trip in a day and slept outdoors
en route. Residents of Simizino village of Vladimir region wrote in 1972 that
"simple people like us have to travel to Vladimir even to buy bread, and the
bus stop to go there is a ten kilometer walk from our village."[27]

Because of distance, by the 1970s there emerged "traveling stores," or trucks
that brought basic goods to faraway villages for sale. The choices were few
and far between, the trucks were not efficient and often did not make it to
their final destination.[28] But nonetheless, they were the only lifeline for many
rural women in places that had no grocery shops where the closest stores were
dozens of kilometers away. To fill in the gap, some professionals who went
from place to place also carried out small-scale trading. Because house visits
were routine and the most common form of medical care, medical professionals sold basic drugs when they traveled from village to village and
functioned as miniature pharmacies. Mailmen often sold small items such as
matches, salt, and of course, newspapers and, upon request, books. But there
were no clear delineations, and a nurse could also bring one's pension from
a local center, some food, medication, and newspapers.[29]

Large purchases were traditionally made in cities, while smaller things were
expected to be purchased locally. But some items were hard to come by in
regional stores. One woman complained that the residents of Kislovo village
of Riazan region could not find warm winter hats in any of the stores located
in nearby urban centers. Equally problematic were dishes, mugs, buckets, and
jugs.[30] Even travel to large cities did not guarantee villagers that they would
find what they needed for everyday consumption. Indeed, Moscow stores sold
a large variety of goods, and resale stores even boasted imported items from

Western Europe, Japan, and the US. But to most peasants these items were exotics of little use. What they needed and could not find were cotton stockings, rubber boots, padded jackets, and fur hats with ear flaps. None of these items were available in the city, mostly because there was no demand for them among "sophisticated urbanites." "We need felt boots, padded jackets, and hats," wrote residents of Mordovian village to a Moscow-based journal in 1969,

> But in our stores, jackets that most tractor drivers and other collective farm workers need to work in are not available for sale for decades . . . Where are factory-made felt boots? They show up occasionally in local stores, but only of poor quality wool . . . In the last 2 years, even fur hats disappeared![31]

Sometimes even soap, salt, matches, and kerosene became deficit items in the countryside. "Where do we find matches?" wrote a woman from a village in Gorkii region. "We are back to relying on friction and stones as our tools to start a fire."[32] Another woman from Kirov region complained that "We do not have soap for half a year now. We bathe and wash ourselves with *shchelok*" (a mix of ashes and water).[33]

The biggest problem was the incompatibility of items sent to villages' stores according to Moscow (and thus urban!) directives *and* the reality of rural life. Women complained that they had no problems finding nylon stockings yet they had no use for them because nylon is too fragile and cold for agricultural work in rubber boots. They needed cotton stockings instead. The same went for almost all items. Underwear was made of synthetic materials, and, although visually appealing, it did not provide the same comfort as knee-long cotton underwear, which was hard to find in stores. Shoes with heels were of no use; what rural women needed were rubber boots, felt boots, and faux leather boots.[34] Instead of regular hats, they needed headscarves that could be tied at the back to cover up hair and keep it out of the way when milking or gardening.[35] Fancy sweaters were equally out of place. They cost three or four times more than loose-fitting simple cotton or wool tops, which women "needed for work and [appreciated] that they were warm, cheap, and easy to keep clean."[36]

In their spare time, rural women often made their own clothing or repaired what they already had. A sewing machine was a rarity in the immediate post-war years, and a family that owned one was considered prosperous. A bride with a sewing machine was considered well-prepared for a family life, and little by little, the sewing machine became the basic necessity of all rural families.[37] Although some women sewed their own dresses, skirts or blouses, most work was invested into repairing existing clothing. Children's clothing was alomst never purchased but tailored by downsizing parents' and older sibling's outfits. Outerwear was commonly "turned inside out," i.e. most materials used for outerwear were double-sided and once one side wore out, a woman could re-stitch a coat or a jacket and turn it literally inside out for a fresher look. Getting "unfashionable" hand-downs from urban relatives was also common. Nonetheless, clothing that was home-spun and homemade from scratch became rare in the second half of the twentieth century, and completely disappeared by the 1970s and 1980s. Women made do with clothing that was manufactured, even if it required extensive tailoring or had been used for many decades.[38]

Repair services, along with the service sector in general, were largely unavailable to women, or only available sporadically. Dry cleaning and repair shops were not available everywhere and were often located a great distance away. Poor roads often made even a short distance impassable. By the late 1980s some 50,000 businesses in the service sector were registered in the Soviet countryside, but they reached only an insignificant portion of rural dwellers.[39] One typical letter sent to a Moscow-based women's magazine reported that:

> our refrigerator worked for six months and then broke down. [The] repair shop that services our warranty is located in a town X eight kilometers away from our village. Despite the distance and poor road, we took the fridge there. As it turned out, [we did it] for nothing, because we were told to take it to town Y which is 120 kilometers away. After much pain and suffering, we finally made it there, and as it turned out there, we had to make an appointment [and could not be seen] but the repairman could come to our village. We made an appointment for the repairman to come, took our fridge back, and have been waiting for the repairman for six months now.[40]

Even in the late 1980s, 20 percent of families were significantly dissatisfied with the stores, repairs shops for TVs and household appliances, and other services available to them (some had no services at all and thus nothing to complain about).[41] At least a third of all rural women constantly faced an added stress because of the non-availability of such services.[42]

Of course, rural life was not static throughout the Soviet years, and the comforts and conveniences that were available to rural dwellers changed and improved in the last two decades of the Soviet regime. No matter how abysmally poorly the Soviet system was performing, on average (although not universally), domestic chores in the countryside became relatively easier, and more conveniences became available. In 1970, half of all rural families had radio receivers, and a third had television sets. One in 10 families owned a refrigerator.[43] But by the late 1980s, most families had television sets and three-quarters had radio receivers. Over 60 percent had refrigerators, and 25 percent even had vacuum cleaners.[44] Although these numbers were only a half compared to that of urban residents in the late 1980s, nonetheless they were indicative of the broader changes in the countryside.[45] The demands of the domestic chores and the time invested in them went down as household appliances became more prevalent. The widespread system of daycare facilities that finally reached the Soviet village was also instrumental in freeing women for other activities.[46]

Ironically, some appliances were resented by peasant women themselves. It became a prevalent notion to reject any use of washing machines, even when they became available and accessible to rural women. Some women argued that it was a waste of money because "it is not a big deal to wash our laundry by hand." Others continued to use washboards because they believed that women can wash the laundry cleaner and make it last longer, while washing machines did nothing but damaged and tore clothing.[47]

Although a rare occasion, rural residents nonetheless had leisure time as well, and enjoyed watching television, going to the cinema, to a local club, or pursuing their hobbies. Clubs were popular in the Soviet days, and they attracted all those who wanted to watch a movie (clubs also functioned as movie houses), or wanted to play chess and checkers; read a newspaper or a book; pursue embroidery and various crafts; learn how to dance; go to a dance night; or participate in a folk ensemble or an amateur chorus. The absence or presence of such clubs often shaped the experience of a rural life, as having a popular cultural venue such as this meant having some semblance of social life. Villagers who did not have clubs were faced with a prospect of spending all their spare time at home or at their neighbors' houses, and rural residents often complained that the great distance separating them from a club meant in fact a separation from "real life."[48] The number of those who enjoyed clubs and those who did not was nearly equal, as roughly 50 percent of villages had one and about 50 percent did not.[49] On many occasions people complained that:

> in our country, there is a grandiose building project to have clubs, houses of culture, libraries, and stadiums [available to all] so that Soviet young-sters can go there after school or after a workday and have quality time, maybe relax, dance, show off their nice outfits and their talents in the amateur clubs and in sports, and learn more. The construction of a club in our village has been an ongoing project for 3 years. But at the moment, all construction stopped. All materials assembled for this club were relegated to building a new cow barn . . . In response to our inquiries, we were told that . . . we could socialize without a club as we had done previously.[50]

Nonetheless, most regions saw new clubs open on a regular basis to satisfy the need for quality time among rural residents.[51]

Reading was a common choice for village intelligentsia, but ordinary collective farmworkers read magazines and newspapers as well. Books about war experiences, adventure literature, and even some historical works were also popular. Most households owned some books.[52] Television became remarkably popular once it reached the countryside. Choices were limited because broadcasting was limited as well, but films, concerts, sports, and news were always welcomed in the countryside.[53] Cinema was easily the most popular pastime. Going to the movies in a village was an experience quite apart from that in a city. In a village, all members of the audience knew each other, and after the movie was over, people usually tended to stay for a long time discussing the movie's pros and cons.[54]

Also common was attending the classes for basic literacy that were available in the evening for those who lacked a rudimentary education. Various campaigns were aimed at eradicating illiteracy in the countryside, with the largest group of illiterate people being older women.[55] Although some illiteracy was reported on June 1, 1959, the census of 1959, conducted only months later, reported that illiteracy was fully eradicated in the Soviet Union. By the mid-1970s, both villages and cities were subject to mandatory school education, and by the 1980s most peasants—both men and women—received school diplomas or vocational training, or were at least literate.

18 The special environment of the village life

Life in the countryside, with all its challenges, continued to allure peasants with features that were unknown in larger urban centers. Some dwellers appreciated that rural life was a mix of old and new, tradition and innovation, age-old habits and scientific advancements. Others loved the close proximity of nature. Many rural residents commented that they felt "exuberant happiness" to be so close to nature. "When getting to work, there are sunflowers bright as dawn on one side, and grain high and strong on the other side . . . and you go and see that there is real nature and real happiness there."[1] Nature also dictated the pace of life as most agricultural work was seasonal. While the work was heavy and demanding during the planting and harvesting seasons, there was none of the chaos and madness of urban life, and the speed of life slowed down significantly during the winter months.

However, the most important and unique feature of rural life is the sense of a closely-knit community, which is shaped predominantly by close relations. The village life was, and continues to be, less anonymous and less private than urban life, and is subject to closer scrutiny by relatives, neighbors, and friends. In cities, such a degree of indirect control over each other's lives was often perceived as intimidating, as it took away people's privacy. But villagers thought otherwise; the proximity and close engagement of people who have known each other since birth was valued rather than resented. If the urban life was divided into *we* (the family) and *they* (neighbors, co-workers, others) categories, then in the countryside everyone was "we" and there was no "they." The sense of belonging to the community was strong and very pronounced, and villagers often felt that even though they had their country as their motherland, their village was their "smaller motherland" where they belonged.[2]

Because village life was so enclosed, every personal event or a family occasion, no matter how minor, immediately became a subject of discussion across the entire community. On the streets and by the well, women often gathered to chat about their friends and neighbors, to exchange "news" of scandals and cheatings, and spread rumors about things on which they had only second hand information. News spread like wildfire in the village, and nothing could stay out of the public eye for long. Men typically learned of

Figure 18.1 Collective farmworkers from a social *artel* are dancing during their break in 1936, Russia, Crimea.

Source: © Courtesy of the Russian State Documentary Film and Photo Archive, Krasnogorsk, Russia

such "news" from their relatives, but it was unacceptable for a man to express a different opinion than that of his wife and to take the other side. Men typically got together to talk about politics or work and not family matters. Women, however, primarily discussed family matters, children, and household work. The discussion of the private lives of neighbors, and adding detail to rumors without verifiable evidence, was a common and expected occurance.[3] Rumors were not welcomed per se, but they formed an integral part of village life and played a crucial role in shaping public opinion about individual members of the community. Rumors could break or make a life, and a family's reputation could easily be destroyed by them. The Institute for Social Studies of the Russian Academy of Sciences conducted research, which demonstrated that rumors shape the so-called "public mentality." As it turned out, they also contain more truth than expected, usually about 90 percent with only 10 percent of added and unjustifiable myths.[4]

Rumors were also crucial at times of rapid political and social changes or catastrophes. For example, Stalin's death prompted speculations about what was going to come next and was perceived as a personal tragedy by many people.[5] Without any evidence, people rumored about an increase in taxation and a monetary reform. As a consequence, many rural residents took the savings they kept at home and deposited them into a state bank, thus increasing the overall deposits by two to three times in a single week in 1953.[6]

The Soviet system also allowed its people to vent their personal envy or frustration by writing denunciations to party cells and reporting "abuses of the socialist system." The Central Committee of the Communist Part of the Soviet Union registered thousands of letters from villagers who denounced their neighbors for being wealthier than they were, or more successful overall. Sometimes such "compromising materials" were accumulated for years and carefully recorded. Some letters, such as one from an ordinary collective farmworker written in 1959, provided financial details of all collective farm transactions, including the farm's expenditure for machinery and repairs. The goal was to demonstrate that the funds were abused and use inappropriately, but the systematic approach to collecting information, and the details known to an ordinary worker, are both stunning.[7] The stream of letters started to dwindle only after the demise of the Soviet Union, when 70 percent of men and 62 percent of women began to believe that writing letters was pointless because no one paid any attention to them.[8]

Research conducted after the demise of the Soviet Union also confirmed the preliminary findings of the Soviet days—even though men were revered as heads of their household, women exercised significant informal control over certain spheres of life. Thus, because women were held single-handedly responsible for upkeep in the house and for all chores, they claimed significant control over family finances. Only 2 percent of all men had full control of family finances, whereas 27 percent of women made financial decisions by themselves. Another 26 percent had the final say when it came to finances, even if they consulted with their husbands beforehand. A total of 60 percent of respondents of both sexes reported that husbands did not make significant decisions about money without first consulting with their wives.[9]

The post-Soviet transformation was also a challenging and psychologically demanding period in people's life, which saw the disintegration of a known lifestyle and a change to a new system. But Russian women turned out to be better able to withstand the hardships. Typically, according to research sponsored by the World Bank, at the time of social and political transformation women suffer from stress-related depressions more than men; on average 30 percent of women suffer, compared to 13 percent of men.[10] Yet similar research conducted in Russia demonstrated that as of 2001, 53 percent of those who suffered from stress-related depression were male and 47 percent were female.[11] Women invest insufficient time in taking care of their health, but they work hard and succeed at taking care of their families and remaining sane in the process.[12]

19 Protection of childhood and motherhood in the countryside

The roots of the social security in Russia go back to the late nineteenth century when the tsarist government inquired into the nature of social security and implemented basic provisions for the sick, elderly, and invalids, as well as measures to promote, protect, and provide for motherhood. But the expansion of these legislative provisions would take place after the Revolution of 1917 when the social programs, which were extended to the majority of the population, would surpass similar programs in most other countries of the world. To the present day, the system of social security, including provisions for motherhood and childhood that exist in Russia, is one of the best in the world.

In October 1917 the Bolshevik government announced the need for a change to the state benefits system for workers who were in need. It created the People's Commissariat (Narkomat) for the State Oversight of the RSFSR. Its goals were to provide social security to workers during temporary unemployment; protection and state provisions for motherhood and childhood; social benefits for war veterans and their families; and social welfare programs for elderly and children.[1] In 1918 the Narkomat for the State Oversight was renamed the Narkomat for Social Security, and its function for protection of motherhood was transferred to the newly created Commissariat for Health (NKZ). The system of a state insurance, that aimed to protect in times of temporary disability and other such social disturbing instances, was replaced with a free social provision for people and a universal availability of various free services. The same year the NKZ issued a Decree on the Social Security available to workers, which set the guidelines for a welfare system in cases of short-term disabilities, unemployment, maternity leave (as well as childcare provisions), pensions for invalids, and accessible medical care for all.[2]

In 1933, the responsibility for a social security system was relegated to labor unions, and the main cost of financing various programs was firmly established as part of the state budget. At the governmental and administrative levels, social security became subsidy-based, and the basic structure of social security and its financial base remained unchanged into the 1990s.

Various specific decrees and measures went to establish the newly minted Soviet Union as one of the states in the world with the best social security systems. For example, the Decree "On the Insurance in the Case of Illness" from December 22, 1917, provided for monetary payments in the case of illness, childbirth, death of the employed head of the household; various free medical services such as access to emergency care and in- and out-patient procedures; household help for the disabled; childbirth facilities; a free room and board for patients who required hospital stays; and even free *sanatoriy*, a health-improvement and rest facility for extended stays of 21–24 days.[3]

The problem of orphans and abandoned children attracted significant attention from the state, but the government attempted to provide social services to *all* children who needed state support. All children were assured of a safe and sound place in orphanages or extended daycare facilities. More than 1.2 million children were under direct state supervision in 1920. Soviet Russia further reformed the system of education and after-school programs, now under the supervision of the People's Commissariat for Education. The cost of childcare facilities for all was minimal and by law could not exceed 6 percent of a parent's paycheck.[4] But these are only a few examples of nearly 100 decrees and orders in relation to social security issued by the various Commissariats during 1917–22. All these acts, along with nearly 1,500 newly opened centers that provided various services to the people, laid the foundation of the Soviet system of social security.

However, all measures of social security were only available to the urban population, and to a lesser degree to men who migrated to cities in search of better wages. But the entire rural population—and thus the majority of Soviet people at the time—missed out on these benefits. Basic social security pro-visions were supervised and implemented by the so-called organizations of Peasants' Mutual Assistance, created in 1924 to offer subsidies and benefits to the families of military personnel, invalids, and the poorest of the poor among the peasants.[5] In 1929, in the midst of the collectivization campaign, the government also issued a decree that authorized "the support of persons and households who suffered from the kulak exploitation." The decree once again affected the poorest strata of the peasantry by providing a one-time subsidy, extending medical insurance equal to urban works, and ordering priority placement for children entering various childcare facilities.[6]

Collectivization, devastating to many households and even agricultural production itself, also reorganized the social security system in the country-side. Most programs and services became collective-farm-based; were treated as the responsibility of each collective farm; and also implied a decree of differentiation from one collective farm to another. But in the sphere of child-hood, collective farms became responsible for organizing childcare facilities for children of two age groups—newborn up to 3 years old, and 3 years old up to school age (7 years old). Collective farms also oversaw the building of playgrounds for these children to use while they were not in daycare.

Maternity leaves for women in the collective farms were also highly indicative of the massive gap that separated urban and rural Soviet populations. As of 1935 rural women were entitled to a two-month maternity leave and

50 percent of their pay during the leave, but urban women had twice as long and twice as much payment. Peasant women also faced significant informal opposition and mild harassment from their collective farm leaders if these new mothers wanted to leave the workplace during the sowing and harvesting seasons.[7] In the 1930s, collective farms also established funds for elderly and disabled workers.

In the Constitution of 1936, article 122, women were assured equality to men in all spheres of life, including employment, work compensation, social security, and rights to education. The article also reaffirmed provisions for the protection of motherhood and childhood, such as the right to a maternity leave and daycare facilities. But World War II interrupted the implementation of these blueprints of equality, not the least because they became irrelevant in a war-torn and asset-stripped country.

The government did not change its emphasis on the protection of motherhood in the late-war and immediate post-war period. By 1945, the government continued to provide pregnant women and breastfeeding mothers with extra ration cards, and authorized the distribution of the so-called "children's dowry," which consisted of cotton cloth for diapers, baby blankets and clothing such as diaper shirts. The Soviet Office of the Prosecutor (*Prokuratura*) was also held responsible to overseeing such distribution. The sweeping campaign of 1945–46 undertaken by the Office of the Prosecutor, "revealed the facts of breaking the law" and "the facts of lowering subsidies." In the Moscow region, women were only able to obtain one-third to the total allowed on the ration cards, while in Gorkiy and Vologda regions only a fifth of all women who were entitled to extra milk and dairy products managed to secure this extra provision. Voronezh, Stalingrad, and Sverdlovsk regions, as well as Tataria, suffered from the same shortcomings, according to the findings of the Prosecutor. In Chelyabinsk region, however, in several towns "pregnant women and breastfeeding mothers did not get any additional food rations."[8] The number of children routinely exceeded the reported number of "baby dowries" by a factor of two, while those women who managed to receive their norm often waited a year or more, long enough for a child to outgrow diapers and sleepsuits.[9] The same campaign revealed cases of "minor theft," "improper usage," "wastage" of extra provisions, or an overall "lack of necessary attention and control" for the matters of providing for new mothers and their newborns.[10] As a consequence, the responsibilities for this "necessary attention and control" of motherhood were relegated to the Ministry of Trade and Ministry of Health, which in tandem had to secure food and basic consumer goods for women with small children at the time when the rest of the country still lived by ration cards. The system of rations cards was mostly canceled after 1947 and the need for additional ration cards for mothers ended with it. But a special privileged access of expecting mothers to basic consumer goods for childcare (baby soaps, diapers, cotton cloth) remained intact for the duration of the Soviet state.

Even though the country as a whole experienced what many perceived as a miraculous recovery, the people of the countryside were largely left out from the state security system. First real benefits reached rural Soviet populations only in the second half of the 1960s. Only then were rural residents offered benefits on a par with urban Soviet citizens. On July 15, 1964, the Supreme Soviet of the Soviet Union passed the Law on Pensions and Subsidies for Collective Farmers, which became effective on January 1, 1965. According to this new regulation, farmers were finally granted the social security provisions available to other citizens of the country, including pensions upon reaching retirement age; subsidies and payments in case of a death of a head of the household if he was the single provider; and disability payments and maternity leaves for women who worked in the collective farms. It was not until 1965 that rural residents received rights equal to the rest of the country and became full citizens with all rights extended to them.[11]

In 1969, a new Fund for Social Security of Collective Farmers was established. It became a source of financing and distributing such social provisions as short-term disability benefits, free passes to sanatoriums, summer camps, and vacation houses. By the late 1960s, the system of social security and all the benefits it provided in a socialist state became universally available to all people, regardless of their place of residence. In 1969, the Soviet government outlined basic provisions for health insurance and medical care that was available to its people, and in 1970 it outlined the basic provisions on the security and protection of labor and on equal pay for equal work. The Constitution of 1977 fully incorporated these provisions and created an extensive system of social security and free medical care. The Constitution stipulated that all citizens had equal rights to social benefits; that there should be no gender discrimination, and men and women had equal rights (article 35); that all citizens had a right to proper medical care (article 42) and a right to a proper pay for their labor (article 43); and finally that the government protected the rights of families, motherhood and childhood (article 53).

In the Soviet Union, all medical services were available to its citizens free of charge, and the medical care system was indeed comprehensive. It also emphasized preventive care (and thus sanatoriums, rest, and vacation houses were common and widespread) and extensive rehabilitation services with prolonged physical therapy for those who needed it. Its emergency care was also available free of charge, and a dispatcher sent an ambulance to a patient even for minor complaints such as a low-grade fever or an upset stomach. The World Health Organization evaluated various models of medical care across countries and concluded in 1978 that the Soviet system of medical care, and especially its emergency response component, was the best in the world. Subsequently, the WHO recommended the system to be adopted by other countries.[12]

Post-Soviet reforms affected the medical care system as well but kept its basic principle of complimentary care. The new Constitution of 1993 outlined that all citizens of the Russian Federation "have a right to free medical care in the state medical facilities." The government allowed people to realize this right by providing mandatory medical insurance policies to all citizens of Russia, largely free of charge. Through this insurance system, all citizens have equal rights and excess to medical care regardless of their age, social status, ethnic background, or place of residence. All employed citizens receive their insurance through their employers, while unemployed citizens and those who are not of working age receive their insurance through specialized social security offices. The insurance and all its fiscal affairs are federally regulated and are under a direct control and supervision by the government.

The insurance policy covers the following services: emergency care; outpatient services and professional consultations; hospitalization; prenatal and childbirth care; physical therapy and all need rehabilitation services, with hospitalization if needed, including those services offered to children and/or in specialized sanatoriums; and rehabilitation services on an outpatient basis.[13]

Expectant mothers are protected in the workplace, and any work-related responsibilities that could possibly endanger their pregnancy must be modified to accommodate the psychological and physiological needs while expecting. All prenatal care and childbirth are fully subsidized and entail no out-of-pocket expenses for parents. Also, women are granted the following rights for childcare: maternity leaves of up to 3 years; a one-time monetary subsidy at childbirth; full wages if employed or state subsidies if unemployed up to when a child is 18 months; state child-support payments to single mothers until the child reaches the age of 16; and monthly state-financed child-support payments to children of military personnel until children turn 16. The government also guarantees sufficient nutrition to expecting mothers and children under the age of 3 years, through the means of state subsidies if needed.

To keep up with world medical standards, by the late 1980s and early 1990s new centers of Planned Parenthood were established across the country, and by the late 1990s, over 1,000 such centers offered medical assistance to families who were not able to conceive on their own. These centers also had medical professionals who specialized in problematic pregnancies and hereditary diseases, and extended medical care to those expecting mothers and newborns who could not receive such specialized care at regular maternity wards.

The idea that prenatal care and childbirth assistance was of crucial importance and had to be available to all, even rural residents, goes back to the times of Peter the Great. In the early eighteenth century Peter created training facilities in St Petersburg, Moscow, and Kronstadt for future medical professionals of "general medical and surgical practice." First schools that specialized in a "woman's business," to use the term of the time, opened their doors to future professional midwives in the second half of the eighteenth century. The first such school welcomed 20 students into a program that offered 6 years of medical education and training. In 1797, the first Midwifery Institution (*povival'nyi institut*) was founded in St Petersburg. By 1860s, several similar schools were operational across the country, and by the early twentieth century 40 such "midwifery institutes" offered professional training to hundreds of medical care professionals.

In pre-revolutionary Russia, midwives were classified into two categories. The first category included midwives who received their training at the Midwifery Institute and took medical examinations. They had the right to practice across Russia with no restrictions and were usually highly skilled medical professionals. The second category consisted of midwives who were trained at one of the smaller, local schools and were allowed to practice only in the countryside in smaller villages. But both were required to follow the Medical Code, which set the following guidelines: a midwife should be available at any time, be that day or night, and should not leave a woman until there is no longer any need for a midwife; a midwife should be caring and patient; and a midwife is expected to call a doctor if she judges that the situation has became of concern for the mother and or child's health and lives. Yet there was a drastic shortage of midwives in the countryside, and at the beginning of the twentieth century, only 5 percent of all births in the countryside were supervised by a skilled midwife. The main reason was the distance that separated the expecting mother and the midwife, and the lack of suitable and accessible carriages or sleighs to transport either the woman or the midwife.[14]

The rural life dictated its own rules, and the proverb "a woman gives birth where she is standing" reflected that reality well. Peasants happened to give birth whether they were at the time, in the fields, in the woods, or in an animal barn. Also some peasant women did not trust skilled midwives, mostly because women who received professional medical training were of different social standing, and ordinary peasant women complained that such midwives did not understand "the real peasant needs and realities" because they were "refined." Instead, many peasant women preferred to call in a local woman who performed functions of a midwife. Usually, such a woman was a widow, usually older and beyond child-bearing age, someone who had seen many births before, and was from the same village, and thus the same mindset and the same community as the expectant mother. These "midwives" also took it upon themselves to help around the house in the first few days after the birth— they cooked, cleaned, and supervised older children while the mother and newborn were recovering. Unlike professional midwives, these women received their compensation not from parents but from guests who came to a child's Christening ceremony, and often the payments were in kind. Some women were rather skilled as they had supervised many births, but they rarely called for professional medical help in cases of complications and relied on prayers instead of medical expertise to address problems.

There were also traditional rituals that these older women performed to "facilitate" birth. Common among them were asking a woman to blow air into a bottle; to scare a woman; to unexpectedly pour cold water on her; and to undress her fully so that she could "earn" by labor her things back. It was also believed that the fewer people knew about birthing, the better it would go. Hence women were often taken to bathhouses or animal barns where only the expecting mother and one or two women were present. During harvests, women were expected to go back to work on the third day after giving birth, and at this time "midwives" often "helped" women recover by making them stand on all fours and lifting their feet up, and shaking women, practically upside down, "to get everything in there faster."[15] Unsurprisingly, the death rate associated with childbirth was high; approximately 30,000 women died in childbirth annually in Russia, and the life expectancy for women was 33 years. The situation was not much better for children. Even though the first Children's Hospital was opened in St Petersburg in 1834, it serviced only a small number of children. On average, 40 percent of all children died before reaching the age of 5, and in some regions the death rate exceeded 50 percent.[16]

Hence the newly established Soviet government made the protection of childhood and motherhood its top priority. In November 1917, the People's Commissariats for State Supervision (of motherhood and childhood) was founded, and in January 1918, a special Committee was formed to oversee the urgent need of new mothers and to propose measures to address such needs. These functions were relegated to the People's Commissariat for Healthcare once it was created in the summer of 1918. Despite a widespread famine and the devastation of the war, in 1918 the new government opened 46 extended care and orphanages, 66 maternity wards, 59 prenatal care centers, 47 centers for (re)distribution of breastmilk and infant-appropriate meals, and 409 daycare centers for newborns through to the age of 3 years. Furthermore, the Soviet government issued decrees on "Protection of Motherhood and Childhood" (1918); "On Creating a Soviet for the Protection of Childhood" (1919); and "On Organizing Children's Feeding" (1919). In 1922, the State Institute of Motherhood and Infancy was founded, which was later, in the 1940s, renamed into the Institute of Pediatrics. At the same time there was a new Institute of Obstetrics and Gynecology. Already in 1925, over 1,000 centers were created exclusively for the needs of expecting and new mothers and their newborns. All these measures brought immediate results; in 1925 the death rate among children under the age of 5 went down to 20 percent, and it continued to decline for years thereafter.

World War II gave a new direction to the work of various committees that oversaw the question of childhood. Their top priority was to ensure the evacuation of children to safe zones and provision of basic medical care to these children in their new places of residence. At times, the aims of ensuring satisfactory living conditions included basic hygiene standards and simply supplying evacuated children with clean, drinkable water. Although the conditions in the war-torn country did not allow for a perfect implementation of such plans, overall the effort to evacuate and save children was perceived at the time, and later evaluated, as satisfactory or even good.

In the post-war decades, the main places for medical care were pediatric centers, which all children were assigned to based on their place of residence. All children were at least nominally under the supervision of a pediatrician. Moreover, by the late 1950s there were already 7,000 prenatal care centers, and all women in towns and cities gave birth in specialized maternity wards staffed by medical professionals.

The later regulations of the 1970s, especially the Legal Code of 1970, assured that all women were entitled to free prenatal care; free medical assistance during birth and postpartum; free pediatric care for children; unlimited paid leave to care for sick children; free and equal hospitalization; free preventive medical treatment for themselves and their children; and universal access to free and nearly-free daycare facilities for children before they reach school age (which was seven in the Soviet Union). All work related to motherhood and childhood was supervised by the Ministry of Health of the USSR, which included a specialized Department for Medical Help and Preventive Medicine.[17]

It was not possible for remote villages to have professional medical help universally available to them, although considerable effort was invested in bringing traveling teams of obstetricians and gynecologists, for prenatal screenings and health checks, to faraway places. In addition to maternity centers, there were always maternity wards and departments in general hospitals. Yet predictably, rural Russia lagged behind the urban centers in providing universal health care to mothers and children. Even in the 1950s, only 12 percent of all rural women gave birth under the supervision of a medical professional, and the rest gave birth either unassisted or with the help of female relatives or "experienced women." One woman, a record-setter for milking cows, remembered:

> I gave birth to all my eight children next to cows. When I am on my way to do morning milking, the road would get bumpy and I get too shaken from it. So by the time I get to my cow barn, I already have a baby on my hands; [this is] a common thing. Sometimes we both gave birth at the same time; a cow was having her calf in a barn as I was giving birth to my baby next to it or in the house. I stay put for a couple of days and then go back to my cows as usual.[18]

For the duration of the Soviet era, the main obstacle to getting quality care in the countryside was the distance to the nearest maternity ward. It was common to travel 10 kilometers to the nearest medical center, and under such circumstances most women ignored prenatal care altogether. By the 1950s and 1960s, horses in private possession also became a rarity, and the collective farm directors seldom allowed using a horse to get a woman to a maternity ward. One woman, M. Zolotareva, remembered that she routinely traveled more than 30 kilometers on a horse-drawn cart to transport sacks of grain. Yet when she requested to use the same cart and the same horse to go to a hospital that was 30 kilometers away to deliver her baby there, she was denied her request. As a result, she suffered numerous complications during her unsupervised birth and remained crippled for life.[19] Some women also saw it as inappropriate to demand much attention to childbirth, which they considered the private business of a woman. But although the change was slow and lagged behind the city, the number of supervised births in the countryside increased over time. By the mid-1960s, enough beds were available in the Russian Federation to oversee 118,000 births at the same time.[20]

Russian villages experienced their most dramatic change in the 1980s. Specifically for rural residents, the government financed an additional 10,000 beds in children's hospitals; additional pediatric offices to oversee 30,000 visit a day; new maternity wards with 7,300 beds in them, and new prenatal centers for 17,000 visits daily.[21] By the late 1980s, nearly 50,000 centers offered medical care to rural women who were expecting, were in labor, or were new mothers.[22] The number of villages that had no obstetrical and gynecological care declined as well. In 1986, there were 24 rural medical centers that had no obstetrical and gynecological staff, but only 16 in 1987.[23]

Significant attention was also given to the fact that some maternity wards were in an abysmal condition. In 1988, of 500 inspected wards, one-third required immediate repairs, and 6 percent were beyond repair and had to be demolished. The latter wards lacked both running water and sewage.[24] But even those wards that were considered in decent shape in terms of their structural specifications and buildings often lacked bed linens, medication, and medical equipment.[25]

Another problem was that many women had a neglectful attitude to their own health. Even when medical centers became almost universally available, most women did not take any preventive measures, ignored routine screenings, and saw obstetrical and gynecological staff only if they had a medical condition. Of thousands of women polled in 1990, more than 25 percent of women aged 18–23 had never gone to see an obstetrician or gynecologist, and more than 40 percent of women aged 24–29 "could not recall" when and if they went to see their gynecologist for the last time. The same memory lapse of "not being able to recall the last visit" afflicted 50 percent of women aged 30–40; 57 percent of women aged 31–50; and 76 percent of women aged 51–60.[26] Many women disregarded prenatal care even when they had an easy access to it.[27]

Today (and historically as well), most women in Russia have children earlier than their counterparts in other European countries. The average age at first childbirth in Russia is 23 years old (compared to over 29 years in countries such as the United Kingdom and New Zealand), and the average age for having a second child is 25. Maternal death in childbirth is higher for Russia than most European countries as well, although it has been declining overall. In 1970s there were 105 maternal deaths per 100,000 of births—the number went down to 47 per 100,000 in 1990, and to 37, in 2001. (For comparison, the maternal death rate is 4.7 per 100,000 in Australia, 4.3 in Italy, 3.6 in

Switzerland, and 1.7 in Norway, but 21.8 in Armenia, 23.9 in Ukraine, and 58.2 in Tajikistan).

According to the findings of the Ministry of Health in 1989, 50–70 percent of all pregnant women suffered from complications and needed medical care. Thus, 15 percent suffered from anemia; 4 percent from kidney-related problems; and 4 percent from cardio-vascular problems. Only a half of all births proceeded without complications, and more than a half of all newborns had health problems that required medical attention and treatment in various medical facilities.[28] This data was alarming, but the trend got even worse in post-Soviet times. In 2001, 43 percent of women had anemia, 20 percent had urinary tract problems, and 10 percent had cardio-vascular problems.[29] Moreover, in 1996, only 34 percent of births were without complications, and their proportion went down to 31 percent in 1999 and 32 percent in 2000. Over 40 percent of newborns suffered from medical conditions that they acquired in their first weeks of life, often while still in the maternity ward (an expected length of stay in a maternity ward is five to seven days) and 25 percent of newborns had hereditary diseases. The death rate for infants in their first year in Russia is two to four times higher than in other developed countries.[30]

The financial crisis of the first post-Soviet decade negatively affected the quality and quantity of medical services available to women and children. The availability of medical centers declined, and centers that stayed open needed repairs and better equipment. A total of 12–15 percent of hospitals were in deplorable conditions (beyond repair), and 25 percent needed repairs urgently. Two-thirds of rural clinics and hospitals also lacked even basic medical equipment and medication. Construction of new hospitals became even more problematic as the building of rural hospitals went down five times, and of outpatient clinics seven times.[31] In 1998, one rural resident complained that:

> To get medical care in a hospital, I need to buy all necessary medication [hospitals lacked money to purchase it], and protective rubber gloves [for medical personnel]; I need to bring my own bed linens, and meals are so poor that everyone eats what they get from home.[32]

The majority of Russian citizens evaluate the change in medical services in the 1990s negatively.[33]

To address the problem of health care, the government adopted a new resolution, called a National Project Health, in 2005. Its main goals are greater accessibility and quality of emergency care; better training for medical personnel; shorter waiting times for diagnostics and other treatments at the hospitals; a new arsenal of ambulance cars; and the overall lowering of rates of illnesses and deaths in the population.

On January 1, 2006, the government also introduced a program of Childbirth Certificates. These certificates come with a series of coupons, which a woman receives from the government and later uses in lieu of payment for medical services. Subsequently, a clinic uses these coupons to claim payments from the government. Because a woman has the right to choose any medical facility that she wishes, ideally picking the best place, and because the choices that women make can greatly affect the profitability of each clinic, the system created healthy competition and an interest among medical professionals in offering women the best possible care. The certificate includes a coupon for each visit during gestation; a coupon for any services offered during childbirth in maternity wards and/or prenatal care centers; a coupon for each visit to a pediatrician during a child's first year of life; and a general certificate confirming that a woman received medical care throughout her pregnancy and at childbirth.

The clinics submit these coupons to the federal agencies and receive compensation in addition to their share of the federal budget. Yet the compensation for coupons is granted only in cases of satisfactory completion of all services and of a positive outcome of childbirth. All medical centers with a license to perform obstetrical and gynecological care during pregnancy receive an additional 3,000 rubles (approximately $100) for each patient for which it provides coupons; a childbirth center receives 6,000 rubles; whereas pediatric centers receive 1,000 rubles for each child under the age of one. A Childbirth Certificate is granted to all Russian citizens, as well as aliens and foreign citizens who have legal rights to reside in the Russian Federation at the time of childbirth. Each clinic has the right to use this money to supplement staff's wages, to buy new equipment, or to buy additional medication and medical supplies. In 2006, using such coupons, medical centers received 9.1 billion rubles from the federal budget (approximately $330 million), offering childbirth assistance to 1.3 million women and newborns. In 2007, coupons were turned in for a total of 14.5 billion rubles (approximately $485 million) after treating 1.6 billion women and children.

To further promote childbirth and protect the interests of mothers, Putin's government adopted a program of "Additional measures of the state support for families with children." According to this program all parents receive a right to the so-called maternity capital of 250,000 rubles (approximately $10,000; adjusted for inflation annually) for each child after the first. This measure was also a reflection of the abysmal demographic situation in the country. By 2001, the life expectancy for men declined to 57 years in the countryside and 59 years in urban centers (compared to 72 years for women).[34] Once again, women have become the backbone of rural Russia, mostly because by the retirement age there are simply no men left, and some villages are so depopulated by male residents that they have only women and are known as "widows' villages." In recent years the overall population of Russia has been steadily declining by approximately 0.9 million annually (of a total population of about 145 million in 2000; less for later years). In 2005, 2.3 million Russian citizens died but only 1.5 million of children were born, thus once again reducing the population by 0.8 million. It is also exceedingly rare to find families with many children. As of 2007, 15 percent of all families had two children, 3 percent had three children, and the number of families with more than three children was statistically insignificant.

Women have a right to maternity capital if they gave birth or adopted their second or subsequent child on or after January 1, 2007. Men also have a right to maternity capital if they adopted their second child as a single parent, or were left as single providers for their own child(ren) born on or after January 1, 2007. Parents have a right to the maternity capital no matter where they reside or their financial circumstances, as long as they and their children retain their citizenship. Parents have the right to the maternity capital even if they and/or their children have dual citizenship. However, it can be claimed only once, even if a family had multiple children after 2007. Maternity capital is tax free and can be used in various ways. Primarily, families are allowed to use any portion of it or the entire maternity capital for improving their housing, either as a down payment for a new apartment or a payment on their outstanding mortgage. Families also have a right to use this money to offset the cost of education, either at the college level (then money can be used for children until they reach age 25) or any private lessons or school costs (general public school education is free). Yet parents can start claiming money only when the child turns 3 years old, or 3 years after adoption.

Up to early 2009 nearly 7 percent of all families received a right to maternity capital. Compared to the pre-reform years, childbirth went up by 7.7 percent, reaching the highest birth rate in 15 years. The government estimates that the birth rate will continue to increase, not the least thanks to the maternity capital and a sense of government's protection of the childhood that it gives many young families.

20 Abortions

I am not a machine to produce two children a year; I cannot handle giving
birth anymore.

In pre-revolutionary Russia, abortions were both illegal and condemned by the
Church (spousal infidelity was also frowned upon by the Church). Women who
were discovered to have attempted to get rid of an unwanted pregnancy were
separated from the Church for 5 to 15 years, which was considered a severe
punishment for a religiously fervent Russian rural community. But separation
from the Church was also akin to a public announcement of disloyalty and
carried with it years of public condemnation by the majority of, if not the entirety
of, the rural community. Judicially, any confirmed abortions were equaled
to infanticide and subject to the same punishments, usually an exile and hard
labor for 4 to 10 years. But the system was made imperfect by the fact that few
cases ever made it to courts or were publicized. During the period 1897–1906,
only 76 women were exiled and sent to hard labor for abortions, and in 1910–16
on average 20–50 women were prosecuted annually.[1]

Young women were especially prone to public condemnation and humil-
iation, and some women even committed suicide in fear of provoking such
reaction. Others attempted to hide their condition under loose-fitting garments,
then left the village before childbirth and abandoned their newborns at the
doorsteps of orphanages or nunneries. But more commonly, women attempted
to get rid of unwanted pregnancies when they were still barely noticeable. If
the pregnancy could not be terminated early, then miscarriages were provoked
by tightly binding a woman's belly, then pulling robes around it and placing
heavy weights on top of it. Whenever available, chemicals were used as
well to "improve" chances of a miscarriage, and especially common were
gunpowder, various nitrates, kerosene, cinnabar powder (red mercury sulfide),
and arsenic. Most villages had wise women who performed such services,
although pregnancy was always terminated at a high risk to the mother. Some
rather brutal methods were used as well, for example by piercing the womb
with a heated spear.

After 1917, Soviet powers' attitudes to abortions changed several times. Abortion was legalized on November 18, 1920, and women were given the right to abortion in medical clinics under professional medical supervision. Simultaneously, the Legal Code of 1922 and 1926 made performing an abortion without a medical license a criminal offense. Also punishable under the criminal code were abortions performed by a medical professional under "unreasonable conditions" or when sanitation was improper and poor.[2] Abortions without a woman's consent, or which led to the patient's death were subject to close scrutiny.

Early Soviet statistics and propaganda campaigns claimed that abortions were safe for all women; abortions were even "promoted" as being safer than childbirth. The chances of getting an infection during an abortion were 60 to 120 times less than at the time of childbirth. Nonetheless, despite this cheerful picture presented to the public, 20 percent of women who were treated in licensed medical centers had complications after the termination of pregnancy, and some of these complications led to infertility. Painkillers or anesthetics were routinely denied to women during abortions, and when one woman complained of pain, a doctor replied that this should teach women a lesson (presumably, of how to avoid unwanted pregnancies to begin with). As a matter of fact, anesthetics were not used during abortions until the 1990s, when such services became available in private medical centers.[3]

In 1926, over 103,000 abortions were recorded in the Soviet Union, although this number is likely to be under-representative because statistics were better collected in large clinics and urban centers than in the countryside. Thus, Moscow and Leningrad (St Petersburg) housed only 3.9 percent of the female population but represented 39 percent of all abortions. Nearly 85 percent of all women lived in the countryside, and nearly 85 percent of all abortions were recorded in large centers.

In villages, most abortions were performed for women who were married (84 percent) and those who already had children (85 percent). Financial hardships were commonly cited as the main reason for the termination of pregnancy, and the famines of 1921 and 1924 were matched by higher abortion rates. About a third of all women claimed to have "too many children already" but one in five only cared to hide the fact of pregnancy.[4] Women who were reluctant to make their pregnancy known were also less likely to seek professional medical help. They hoped that by going to a "local woman" they would be able to keep their affairs and resulting pregnancies a secret. Such hopes were more often misplaced than not. In 1926, there were 20,000 confirmed cases of illegal abortions, and in regions such as Ivanovo-Voznesensk, with a sex-ratio imbalance (females greatly outnumbered males), almost 40 percent of abortions carried out were done so illegally. In 1930, of all abortions performed in Smolensk region, half were carried out illegally. These numbers are only known because many women ended up in hospitals with bleeding, infections, and various complications. At the time when abortions were nearly universally available to all women, in Nizhegorodskiy region (where all but 6 percent of women were granted the right to termination in clinics), a third of all beds in hospitals were occupied by women with complications after illegal abortions. Nationwide, 14–15 percent of all hospital beds were taken by women in the same position. Some of these cases were fatal, and newspapers, journals, and magazines never failed to publicize the lethal outcomes of illegal abortions. But even such high risks did not prevent women from seeking them.[5]

In 1930, termination of pregnancy became a paid service, no longer offered free of charge to all women seeking one. The average fee was 50 rubles, at the time when the average wage in cities was 80–100 rubles a month. Predictably, abortions became an expensive solution that few rural women could afford. Equally predictably, the number of illegal abortions that led to complications rose dramatically as well. In the countryside, contraceptive methods were often limited to breastfeeding. Women who wanted to minimize their chances of new pregnancies (although this method did not always work) continued to breastfeed until children were at least 3 years old, but many prolonged weaning until the child was 7 or "became ashamed of eating this way."[6] By the 1930s, up to half of all rural women attempted to limit childbirth by abstention, and at least 40 percent knew of other (unspecified) methods of contraception, but only about 18 percent attempted to use them.[7]

The abortion ban of 1936 dramatically changed the situation. The legislative changes of 1936 placed a ban on abortion but they also made divorce harder to obtain and provided state subsidies to mothers with a large number of children (the subsidies and rewards changed in proportion to how many children a woman had). The ban took into consideration all possible places where an abortion could be performed, and explicitly outlined that abortions were banned in clinics and all other medical facilities, as well as at the houses of medical professionals and women.[8] A year before that, a new law of 1935 made abortion illegal in cases when it was a woman's first pregnancy and/or she had no children yet. In 1936, abortions were legalized only in the cases when pregnancy threatened a woman's life or if a woman had significant hereditary medical conditions and illnesses. In such cases, a medical commission had to establish the grounds for terminating the pregnancy, and the procedure could be done only in authorized clinics or maternity wards. All breaking of these regulations was punishable according to the Criminal Code of the USSR.[9]

The abortion ban was a reflection of the government's desire to improve the demographic situation in the country, even though such a ban departed from earlier proclamations about freedom of choice in a communist society. Many rural women welcomed stricter guidelines for divorce, which they felt benefitted women more than men by offering them increased security in marriage. Only younger people and some professional, career-oriented women in the countryside were displeased with the change.[10] The reaction of the Russian village was more subtle than the response in urban centers. Rumors that such a ban would be established prompted a sharp increase in abortion rates right on the eve of this ban's implementation.[11]

The outcomes of the ban were controversial. The enforcement of the ban resulted in higher birth rates in cities, which increased by 50–100 percent in the largest urban locations.[12] But in the country as a whole, in 1937 the birth rate went up by 18 percent, and already in 1938 the increase was insignificant compared to the year before. The trend of a minimal birth rate increase continued during the rest of the pre-war years (up to 1941 for the Soviet Union).[13] Yet the birth rate remained low compared to, for example, the United States, where women had 43 percent more children annually than women in the Soviet Union.[14]

Simultaneously, the abortion ban prompted many women to seek illegal abortions. The overall number of illegal procedures is impossible to calculate because only women with complications who needed medical attention in state clinics "made it" into statistics. In 1937, 90 percent of all abortions were illegal, and this number only reflected women who were treated in hospitals after unsuccessful abortions. The rate of infanticide went up as well; in 1937, it was twice the rate of 1935.[15] In the countryside women were used to relying on local healers and midwives more than medical professionals anyway, so the ban only enforced this tendency to seek no professional assistance. The state could control those doctors who received wages in state-run hospitals, but it could do little to supervise "wise women" in remote villages in the Soviet countryside.[16]

Most commonly the cases of illegal abortions were revealed only when one of the patients died and the case was taken to court. Not all fatal outcomes were blamed on a practitioner, and many deceased women's relatives preferred to keep things quiet in fear of shame and public condemnation that went hand-in-hand with such procedures. Also, some doctors learned to manipulate the system, and for a fee they issued notes that authorized an abortion because the pregnancy presumably threatened the mother's health. Often, medical professionals filled out paperwork that qualified a woman for an abortion on the grounds that an unwanted pregnancy would have negative implications for her emotional, and thus psychological, well-being. Such doctors felt that they were killing two birds with one stone—they made some extra cash for themselves and also met women's needs.[17]

The People's Commissar for Public Health, G.N. Kaminskii (Minister of Health in present-day rendering) found the abortion ban highly problematic precisely because it prompted women to seek illegal services. He suggested to the Central Committee of the Communist Party that women had to have a right to terminate an unwanted pregnancy for social reasons, not just health considerations. He also proposed to expand the list of illnesses that qualified a woman for an abortion. Kaminskii argued in a top-secret document that the ban "created a social catastrophe that had to be dealt with" and cited that in Smolensk almost 40 percent of all pregnancies were terminated and, even in Moscow, 20 percent of all pregnant women chose not to carry to term. Yet all of his proposals were rejected outright.[18]

Discussion of the abortion ban took on a different dimension in the countryside. The most outspoken were younger women who wanted to allow abortions for those women who already had four or more children. They justified their position by the hardships already entailed in raising four children in the countryside.[19] Only women in highly skilled and professional jobs in the countryside rejected the ban outright and saw it as a limitation of their rights, whereas the majority of older women found the abortion ban as a necessary measure. However, they also saw that the implementation of the ban would increase the child-birth rate, so felt that the state should provide more child subsidies to alleviate the hardships involved in raising many children.[20] Women also asked the government to provide more clothing and shoes for children if the government wanted these women to have more children.[21] But the most discussed issues were the questions of medical services and social support extended to mothers and their children, and how these provisions might be improved because of the abortion ban.[22]

No matter how much women argued for a greater access to medical facilities for terminating an unwanted pregnancy, Stalin's attitude was unaffected. "We need people," proclaimed Stalin in 1936:

> Abortions, which destroy life, are intolerable in our country. A Soviet woman has equal rights with men, yet this does not liberate her from her great and honorable duty, which was assigned to her by nature; she is a mother; she gives life. And this is definitely not a private matter but a matter of great social importance.[23]

Nonetheless, not all women were ready to perform their "great and honorable duty" endlessly. An anonymous research conducted by the Ministry of Health in 1950 revealed that most women sought abortions in cases where they did not have a sense of financial or emotional stability in their lives. Most commonly, women who chose not to have a child did so because their husbands had other families (although not officially registered in a marriage); when they expected a child out of wedlock; and when the relationship between spouses was deteriorating and unstable. Other women cited poor housing, poverty, and demands of agricultural labor. Only a few refused to have a child on a whim, simply because they did not want to have one.[24]

In the post-war years, some medical professionals became sympathetic to women with many children. One mother complained that she could not have her fifth child right after she had four in a row and breastfed all of them. Women with many children could appeal to abortion commissions, and those who did were often allowed to terminate their pregnancies because of a fake medical condition that the commission fabricated for them. Yet the law could not be curtailed in all cases. As a consequence, even in the 1950s, women routinely died of complications from illegal abortions, and in 1952 through to 1954 in Moscow alone 400 women died of this cause.[25] Although once again, the exact numbers are impossible to calculate, even in the late 1940s and early 1950s, it was estimated that abortions performed by medical professionals constituted only 4 percent of all terminated pregnancies in the countryside, where 96 percent of abortions were performed illegally. And the number of miscarriages that were provoked is undeterminable, but it has been estimated at 13 percent of all miscarriages in cities and 21 percent in the countryside.[26]

Prosecution for abortion was another problematic aspect of the ban. Women who fared well after the abortion nearly never revealed to the state officials that they had undergone this procedure, and only significant complications led to further investigations. But even then women routinely refused to name those who performed the service, and 67 percent of all prosecuted abortions were classified as "self-abortions" for the lack of evidence that other parties were involved.[27] The Ministry of Jurisprudence even proposed to the government in desperation that husbands and partners of women be prosecuted alongside their female partners and wives. The Ministry hoped both to attack criminal abortions and to reflect the reality that men were often the main instigators and a driving force behind a woman's decision to end her pregnancy.[28] The measure was never adopted, and prosecution for abortions remained a highly problematic feature of the ban. For all confirmed cases of criminal abortions (and "confirmed" already suggested their small number in relation to all such procedures), only 1 percent were prosecuted according to the law.[29]

Although relatively rare, infanticide nonetheless had its place in the Soviet life. Both Soviet and post-Soviet laws treated infanticide as a premeditated murder under aggravating circumstances, although post-partum depression and mental disorders could serve as mitigating circumstances in such cases. Official Soviet statistics recorded that the rate of infanticide was on average 724 cases a year by mid-1950s, and rising to 835 in 1960, although it declined thereafter. In some years and especially in the 1950s, infanticide was 7.5 percent of all premeditated murders, but by 1990 it stood at 0.8 percent.[30]

Facing a bleak situation with criminal abortions, and likely because of the change of the leadership, the Soviet government lifted the abortion ban in 1955. After this date the procedure could only be carried out in clinics and hospitals, and only by trained medical professionals. The law of 1955 continued to list services performed without a license as a criminal offense.[31] But criminal abortions continued in the countryside. In some remote regions, they still constituted a third of all abortions, and this was largely due to the lack of or poor accessibility to medical facilities. By 1970s, an average woman had two to four abortions in her life, and 15 percent of all women treated abortions as a substitute for birth control and had them several times a year.[32] In the 1970s and 1980s, there were 4.5 to 4.8 million abortions performed annually in the Soviet Union, and by 1985 the government even guaranteed paid leaves of absence to women who underwent the procedure.[33]

The popular reliance on abortions also reflected the poor availability of contraceptives. Contraception was sporadically available in major cities, but it rarely ever reached the countryside. In 1989, only 15 percent of all women had access to contraceptives. Even in the 1990s, and despite the ready availability of contraceptives and educational programs, only one in five women practiced safe sex, while the rest still routinely resorted to abortions.[34] Moreover, abortions still typically carry high risks of complications.[35] In the most recent decades, thousands of women have become infertile as a result of complications, and every year 230–240 women die after the procedure.[36]

Despite the high rate of abortions, thousands of children every year were abandoned by their parents immediately after birth. Whatever the reasons (usually socio-economic or hereditary problems), children who await adoption in orphanages far outnumber families in Russia who want them. In the first decade of the twenty-first century, 180,000 children qualified for adoption annually, but only 4,500 families were looking to welcome an orphan. In an act of desperation, the Russian government allowed foreign citizens to adopt Russian children, and thousands of orphans left Russia to go and live predominantly in the US, Canada, Italy, and Spain.[37] Children who are less fortunate and are not adopted typically do poorly in orphanages where even in the best places they routinely lack proper nutrition, medical care, and even basic necessities. As a consequence, about 20,000 teenagers run away from orphanages annually; a third of all alumni eventually receive sentences for various crimes; one-fifth become homeless; and one-tenth commit suicide.[38] However, there are new centers that are established by the ROC and are usually adjacent to nunneries or monasteries, as well as new home-styled orphan communities for 10–12 children each. In all these new establishments, children live with teachers and caretakers with a typically ratio of 1 teacher to 10 or 12 children. The enthusiasm and interests of highly motivated teachers involved in such centers, and the Christian teachings that these teachers follow, ensure that these children receive on average a higher rate of individual attention and emotional support, and as a consequence do better in life than their counterparts in traditional orphanages.[39]

In the 1990s and to the present day, many families claimed that they wanted to have children, and only 2 percent of all families thought that they did not want children at all.[40] But research consistently demonstrates that the financial well-being of most families in post-Soviet Russia was linked to the number of children the family had. Nearly 70 percent of all families with three children lived below the poverty line, and nearly all families with four or more children became impoverished. The income of families with a single child was 3.3 times higher than that of families with three or more children. As subsequent research demonstrated, the incomes of families with numerous children declined progressively with each consecutive child, mostly due to expenses incurred in child rearing and greater time constraints, which resulted in fewer employment opportunities. Families with two children spent 43 percent of their income of food; with three, 48 percent and with four, 55 percent, even though the quality and quantity of food declined for families with three or more children. Hence families had multiple children not because they were poor; rather, they became poor by having multiple children.[41] Unsurprisingly, such dynamics, which were obvious to most people, became a major deterrent to having more than one child. It also did not help that, in 2001, 62 percent of families believed that their life had become worse compared to the Soviet days, and two-thirds argued that their chances of giving an education or medical care to their children also worsened.[42]

The availability of medical facilities was not an insignificant factor. A nationwide medical evaluation of all children conducted in 2003 registered an increase of 26 percent in illness and disease in newborns compared to 1999; of 22 percent in children under the age of 14; and of 24 percent in children aged 15 to 18.[43] Moreover, children in rural areas fared worse than urban children, and the former lagged behind the latter in health, physical development, and physiological maturity. The factors that explain such poor health data included ecological (for example, poor availability of safe drinking water in villages, as well as poor guidelines for the use of chemicals in the agricultural production); socio-economic (for example, lack of prenatal care as well as poor nutrition for children); and socio-cultural (such as poor general health education and understanding of human physiological needs).[44]

Another significant challenge that many women face is the increase in the number of children born to unmarried mothers.[45] The numbers, of course, varied in previous decades as well, not only in the post-Soviet era. Thus, of all births, 17 percent were to unmarried mothers in the 1950s, 13 percent in the 1960s, and at about 10 percent thereafter.[46] Yet up to 1992, the changes of data coincided with general birth rate statistics; the times when the birth rate peaked, so did the rate of children born to unmarried mothers. But in the 1990s, social norms no longer stigmatized cohabitation, and an ever-growing group of women started to argue that with substantial rates of divorce, legal marriage is meaningless. They see no point in "signing the paperwork" because "What? Like the signature is going to keep him in the family."[47] As a consequence, the proportion of children born to unwed parents rose to 29–30 percent of all births and became, according to social scientists, "commonplace and statistically important."[48]

The most significant group of new mothers who do not register their relationship in the countryside is young women, with a third of them in the 15–19 age group.[49] Less than half of all fathers recognized their children, although the rate changed from 43 percent in 1988 to 47 percent in 2000 and continues to rise. First children constitute two-thirds of all children born out of wedlock in cities, and a half in villages.[50] Many mothers are in stable cohabitating relationships and treat their partners as "common law husbands." In other words, in recent years Russia joined the ranks of those Western European nations where social norms allowed for a high rate of childbirth to couples who lack official registration of their relations.[51]

Conclusions

Agricultural and thus rural problems and concerns have been of paramount importance for the Soviet government and the Soviet society for decades. Aside from the economics of agricultural production, the history of the countryside is immensely relevant to the Soviet narrative because the majority of the Soviet population remained rural up to the 1960s, and a third of the Russian society continues to live in the countryside to the present day. Over half of this rural population is female. Hence, to speak of a Russian village means to speak of a Russian peasant woman, and the failure to realize this gendered reality can be devastating. Whenever the government addresses agrarian questions and appeals to the rural population, it needs to effectively address and engage rural women, who still define the daily life of the village and its economics.

In the historical context of the twentieth century, the Russian village is unique in its perpetual efforts to reinvent itself and restructure its productive potential in a specifically outlined period of time according to governmental decrees. This government-driven transformation of Soviet and immediate post-Soviet agriculture was so "successful" that it resulted in the country's dependence on the import of food. Throughout the twentieth century, the government repeatedly changed its priorities for "the revival of agricultural production," which at various times included increased mechanization, reliance of chemicals, soil reclamation, enlargement of farms, restructuring of collective farms and turning some of them into state farms, and even resettlement of "unpromising" villages. Ironically, all of these reforms were typically expected to be carried out with minimal financial support from the state. Propagandistic slogans such as "I choose a village as my place to live" and "All classes—to the animal wards" became routine, and routinely meaningless. At the end of the day, rural women were the only reliable and unchanging labor force in the countryside, capable of endorsing and implementing state initiatives.

Hence, an equally unique feature of the Russian village is its female nature. The years of the gender equality rhetoric did indeed minimize male/female labor divisions, yet typically resulted in a greater workload for women. Today, when about a half of all agricultural produce is grown in private gardening plots, the work of farming and feeding the people remains a predominantly female—and largely manual—task. Work on a private plot has traditionally been treated as an extension of household chores and thus remained in the women's realm of domestic duties. The revival of the village as a viable agricultural force is possible only by acknowledging the importance of the role that women play in it.

Yet village life is not about agricultural production alone. For many people, it is a lifestyle as well. Urbanization of society at large, as well as depopulation and the aging of the countryside destroy the very foundation of a rural Russian life sung by many great Russian poets and writers. While many people proudly identify Kostroma, Gzhel, and other rural neighborhoods and their crafts as the true character and essence of Russia, the same regions of Kostroma, Uglich, Kalinin, Tula, Vologda, and many others have been recording the disappearance of entire villages for years.

The dramatic transformation of the 1990s for the most part did not make everyday life easier for rural women. If anything, its financial chaos brought a sense of insecurity, and its new demands forced many women into working ever-longer hours. The problems in the countryside were, of course, the same as those shared by the rest of the country in this tumultuous decade, yet they were further exacerbated by the remoteness of many villages and the lack of employment options for rural people. Yet the hopes remain high, the hopes for a new village that would allow its residents to combine the comforts and conveniences of the modern world with a commitment to agricultural production and, for women, with the demands of motherhood and care that they extend to their entire families.

Notes

In the Russian publications referenced below, "M." indicates Moscow and "SPt" indicates St Petersburg.

Introduction

1 Pushkareva, N.L., *Russkaia zhenshchina: istoriia i sovremennost'. Dva veka izucheniia "zhenskoi temy" russkoi i zarubezhnoi naukoi. 1800–2000. Materialy k bibliografii*, M., 2002.
2 Tatarinova, N.I., *Primenenie truda zhenshchin v narodnom khoziaistve USSR*, M., 1979; Shishkan, N.M., *Sotsial'no-ekonomicheskie problemy zhenskogo truda*, M., 1980; Novikova, E.E., *Zhenshchina v razvitom sotsialisticheskom obshchestve*, M., 1985; *Zhenshchina v meniayushchemsia mire*, M., 1992; Roshchin, S.U., *Zaniatost' zhenshchin v perekhodnoi ekonomike Rossii*, 1996; Denisova, L.N., *Zhenshchiny russkikh selenii. Trudovye budni*, M., Mir istorii, 2003.
3 *Proizvodstvennaia deiatel'nost' zhenshchin i sem'ia*, Minsk, 1972; *Sovetskaia zhenshchina: trud, materinstvo, sem'ia*, M., 1987; Mashika, T.V., *Zaniatost' zhenshchin i materinstvo*, M., 1989; *Trud, sem'ia, byt sovetskoi zhenshchiny*, M., 1990; *Zhenshchiny: sem'ia, obchestvo, politika*, Penza, 1999.
4 *Zhenshchina na rabote i doma*, M., 1978; Gruzdeva, E.B., Chertikhina, E.S., *Trud i byt sovetskikh zhenshchin*, M., 1983; Denisova, L.N., *Ischezayushchaia derevnia Rossii: Nechernozem'e v 1960–1980-e gody*, M., Logos, 1996.
5 *Migratsiia sel'skogo naseleniia*, M., 1970; Denisova, L.N., *Nevospolnimye poteri: Krizis kul'tury sela v 60–80-e gody*. Chapter III, M., 1995; Denisova, L.N., *Rural Russia: Economic, Social and Moral Crisis*, New York: Nova Science Publishers, 1995; *Demograficheskii krizis: mekhanizmy preodoleniia*, M., 2002; Demchenko, T.A., *Problemy izucheniia chelovecheskogo kapitala v demograficheskom izmerenii v reformiryemoi Rossii*, M., 2002; *Naselenie Rossii na rubezhe XX–XXI vekov. Problemy i perspektivy*, M., 2002.
6 Kharchev, A.G., *Brak i sem'ia v USSR, 2-e izdanie*, M., 1979; Golod, S.I., *Sem'ia i brak: Istoriko-sotsiologicheskii analiz*, SPt, 1998; Verbitskaia, O.M., *Naselenie rossiiskoi derevni v 1939–1959 gody. Problemy demograficheskogo razvitiia*, M., 2002; *Istoricheskaia ekologiia i istoricheskaia demografiia. Sbornik nauchnykh statei*, M., 2003; Denisova, L.N., *Sud'ba russkoi krest'ianki v XX veke. Brak. Sem'ia. Byt*, M., Rosspen, 2007.
7 *Itogovyi otchet o rabote Pervogo Nezavisimogo zhenskogo foruma. Dubna, 29–31 March 1991 god*, Dubna-M., 1991; Denisova, L.N., "Bab'ia dolia" in *Mentalitet i agrarnoe razvitie Rossii (XIX–XX veka). Materialy Mezhdunarodnoi konferentsii. Moskva, 14–15 June 1994 god*, M., 1996; *Materialy Pervoi Rossiiskoi letnei shkoly po zhenskim i gendernym issledovaniiam. "Valdai-96"*, M., 1997; Aivazova, S.,

Russkie zhenshchiny v labirinte ravnopraviia. Ocherki politicheskoi teorii i istorii. Dokumental'nye materially, M., 1998; Ballaeva, E.A., *Gendernaia ekspertiza zakonodatel'stva RF: reproduktivnye prava zhenshchin v Rossii*, M., 1998; Baskakova, M.E., *Ravnye vozmozhnosti i gendernye stereotypy na rynke truda*, M., 1998; *Zhenshchiny i razvitie: prava, real'nost', perspektivy. Materialy Vserossiiskoi konferentsii po polozheniyu zhenshchin. Moskva. 27–28 May 1998 goda*, M., 1999; *Gendernaya ekspertiza rossiiskogo zakonodatel'stva*, M., 2001.

8 *Russia's Women. Accommodation, Resistance, Transformation*, Berkeley, CA: University of California Press, 1991; *Russian Peasant Women*, New York: Oxford University Press, 1992; Goldman, Wendy Z., *Women, the State and Revolution. Soviet Family Policy and Social Life, 1917–1936*, Cambridge University Press, 1993; Goldman, Wendy Z., *Women at the Gates. Gender and Industry in Stalin's Russia*, Cambridge University Press, 2002; Ransel, David L., *Village Mothers. Three Generations of Change in Russia and Tataria*, Bloomington, IN, 2000; Engel, Barbara Alpern, *Women in Russia, 1700–2000*, Cambridge University Press, 2004; Fitspatrik, Sheila, *Stalinskie krest'iane. Sotsial'naia istoriia v 30-e gody: derevnia*, M., 2001; Staits, Richard, *Zhenskoe osvoboditel'noe dvizhenie v Rossii. Feminizm, nigilizm i bol'shevizm. 1860–1930*, M., 2004.

9 *Antologiia gendernoi teorii*, Minsk, 2000; *Vvedenie v gendernye issledovaniia. Part I. Uchebnoe posobie*, Khar'kov-SPt, 2001; *Vvedenie v gendernye issledovaniia. Part II. Khrestomatiia*, Khar'kov-SPt, 2001; *Teoriia i metodologiia gendernykh issledovanii. Kurs lektsii*, M., 2001.

10 "Konventsiia OON o likvidatsii vsekh form diskriminatsii v otnoshenii zhenshchin" in *Narodnoe obrazovanie*, 1989, no 3; "O pervoocherednykh zadachakh gosudarstvennoi politiki v otnoshenii zhenshchin. Ukaz Prezidenta ot 4 March 1993 goda" in *Rossiiskie vesti*, 1993, March 5 ; *Natsional'naia platforma deistvii po uluchsheniyu polozheniia zhenshchin v Rossiiskoi Federatsii*, M., 1994; *Chetvertaia Vsemirnaya konferentsiia po polozheniyu zhenshchin. Pekin 4–15 September 1995 goda*; *Kontseptsiia uluchsheniia polozheniia zhenshchin v Rossiiskoi Federatsii. Postanovlenie Pravitel'stva RF no 6 ot 8 January 1996 goda*; *Mezhdunarodnye konventsii i deklaratsii o pravakh zhenshchin i detei. Sbornik universal'nykh i regional'nykh mezhdunarodnykh dokumentov*, M., 1998; *Piatyi periodicheskii diklad "O vypolnenii Rossiiskoi Federatsiei Konventsii "O likvidatsii vsekh form diskriminatsii v otnoshenii zhenshchin"*, M., 1998; *OON i prava zhenshchin: istoriia i sovremennost'*, Petrozavodsk, 1999.

11 *Rossiiskii statisticheskii ezhegodnik. 2002. Statisticheskii sbornik*, M., 2002, p. 91; "Itogi Vserossiiskoi perepisi naseleniia 2002 goda (sokrachennyi variant)" in *Rossiiskaia gazeta*, 2004, March 31.

12 *Golosa krest'ian: Sel'skaia Rossiia XX veka v krest'ianskikh memuarakh*, M., 1996.

13 *Selo Viriatino v proshlom i nastoiashchem. Opyt etnograficheskogo izucheniia russkoi kolkhoznoi derevni*, M., 1958; Anokhina, L.A., Shmeleva, M.N., *Kul'tura i but kolkhoznikov Kalininskoi oblasti*, M., 1964; *Russkie: Semeinyi i obchestvennyi byt*, 1989; *Russkie*, M., 1997. Also: *Russkie pesni*, Edition 2, L., 1954; *Russkaia chastushka. Fol'klornyi sbornik*, M., 1993.

1 Women's work

1 Ivnitskii, N.A., "Golod 1932–33 godov: Kto vinovat?" in *Golod 1932–1933 godov*, M., 1995, p. 64.

2 "M.A. Sholokhov—I.V. Stalinu 4 aprelia 1933 goda" in *Sud'by rossiiskogo krest'ianstva*, M., 1996, p. 550.

3 *Tragediia sovetskoi derevni. Kollektivizatsiia i raskulachivanie. Dokumenty i materially. V 5 tomakh. Tom 5. Kniga 1. 1937*, M., Rosspen, 2004, pp. 244, 369, 387, 393; Kniga 2, Dok no 17, 66, 144, 145.

4 Zemskov, V.N., *Spetsposelentsy v SSSR. 1930–1960*, M., Nauka, 2003.
5 *Nash sovremennik*, 1980, no. 3, p. 134.
6 *Rabotayushchie zhenshchiny v usloviiakh perekhoda Rossii k rynku*, M., Institut ekonomiki RAN, 1993, p. 4.
7 Denisova, L.N., *Sud'ba russkoi krest'ianki v XX veke. Brak. Sem'ia. Byt*, M., Rosspen, pp. 395–9, 449–53.
8 Denisova, L.N., *Zhenshchiny russkikh selenii. Trudovye budni*, M., Mir istorii, 2003, p. 323.
9 Denisova, L.N., "Zhizn' po kolkhoznomu ustavu" in *Rezhimnye lyudi v SSSR*, M., Rosspen, 2009, p. 157.
10 Ibid.

2 Unskilled labor in the countryside

1 A Scottish proverb; all epigraphs that used Russian proverbs were replaced with English-language proverbs of identical meaning.
2 *Kommunist*, 1983, no. 5, p. 43; Bridger, Susan, "Soviet Rural Women: Employment and Family Life" in *Russian Peasant Women*, New York: Oxford University Press, 1992, p. 276.
3 Vyltsan, M.A., *Zavershayushchii etap sozdaniia kolkhoznogo stroia (1935–1937 gody)*, M., 1978, p. 148; Manning, R.T., "Zhenshchiny sovetskoi derevni nakanune Vtoroi mirovoi voiny. 1935–40 gody" in *Otechestvennaia istoriia*, 2001, no. 5, p. 95.
4 Manning, R.T., "Women in the Soviet Countryside on the Eve of World War II. 1935–40" in *Russian Peasant Women*, New York: Oxford University Press, 1992, p. 216.
5 One centner is 112 lbs; 1 hectare is 2.47 acres. So the harvest was 56,000 lbs per 2.47 acres, or 22,670 lbs per acre.
6 Stalin, I.V., *Rech' na Pervom Vsesoyuznom soveshchanii stakhanovtsev 17 noiabria 1935 goda*, M., 1937, p. 28.
7 Danilov, B.P., Kim, M.P., Tropkin, N.V. (Eds.), *Sovetskoe krest'ianstvo. Kratkii ocherk istorii (1917–1969 gody)*, M., Izdatel'stvo Politicheskoi literatury, 1970, pp. 272–3.
8 Malukhina, A., "Zhenshchiny v kolkhozakh—bol'shaia sila" in *Sotsialisticheskaia restrukturizatsiia sel'skogo khoziaistva*, 1938, no. 3, March, p. 35.
9 VOANPI, F. 2522. Op. 1. D. 43. L. 5.
10 Ibid.
11 *Rossiisko-britanskoe sotsiologicheskoe obslrdovanie rossiiskikh dereven'. 1980–1990-e gody*.
12 Manning, R.T., "Zhenshchiny sovetskoi derevni nakanune Vtoroi mirovoi voiny." 1935–40 gody, p. 98.
13 Manning, R.T., "Women in the Soviet Countryside on the Eve of World War II. 1935–40," p. 221.
14 Selunskaia, V.M., *Sotsial'naia struktura sovetskogo obchestva: istoriia i sovremennost'*, M., 1987, p. 118.
15 "ia i loshad', ia i byk. Ia i baba i muzhik. Ia i seiu, ia i zhnu, Na sebe drova vozhu."
16 *Pravda*, 1989, 9 March; Bridger, S., "Rural Women and Glasnost" in *Russian Peasant Women*, New York: Oxford University Press, 1992, p. 299.
17 Danilov, V.P., Kim, M.P., Tropkin, N.V. (Eds.), *Sovetskoe krest'ianstvo. Kratkii ocherk istorii (1917–1969 gody)*, pp. 314–15.
18 Denisova, L.N., *Zhizn' po kolkhoznomu ustavu*, p. 158.
19 *Itogi Vsesoyuznoi perepisi naseleniia 1970 goda*. Tom 6, p. 166.

20 *Pravda*, 1989, March 9; Bridger, S., "Rural Women and Glasnost" in *Russian Peasant Women*, New York: Oxford University Press, 1992, p. 299.
21 *Sovetsko-britanskoe sotsiologicheskoe obslrdovanie rossiiskikh dereven' 1980–1990 godov.*
22 Denisova, L.N. *Zhenshchiny russkikh selenii*, p. 154.
23 As quoted in Bridger, Susan, "Soviet Rural Women: Employment and Family Life," *Krest'ianka*, 1980, no. 3, p. 11.
24 As quoted in Bridger, Susan, "Soviet Rural Women: Employment and Family Life," *Krest'ianka*, 1980, no. 4, pp. 14–15.
25 As quoted in Bridger, Susan, "Soviet Rural Women: Employment and Family Life," *Krest'ianka*, 1978, no. 8, p. 11.
26 GAVO, F. 1300. Op. 21. D. 2342. L. 158–61.
27 *Vestnik statistiki*, 1989, no. 1, p. 41.
28 "Ustnye svidetel'stva sel'skoi zhenshchuny Marii Zolotarevoi" in *Demografiia i sotsiologiia*, Edition 15, M., 1996, p. 214.
29 Pankratova, M.G., *Sel'skaia zhenshchina v USSR*, M., 1990, p. 108.
30 *Krest'ianka*, 1981, no. 8, p. 18
31 *Sel'skaia nov'*, 1981, no. 9, p. 19.
32 *Sotsiologicheskie obsledovaniia, provedennye v Vologodskoi, Orlovskoi oblastiakh v 1993–1996 godakh I. Koznovoi.*
33 Vinogradskii, V., Vinogradskaia, O., Nikulin, A., Fadeeva, O., "Istoriia sel'skoi zhenshchiny: sem'ia, khoziaistvo, byudzhet" in *Refleksivnoe krest'ianovedenie. Desiatiletie issledovanii sel'skoi Rossii*, 2002, pp. 234–5.

3 Female mechanics and machine operators

1 In the Russian language, a single word *mekhanizator* was used to describe the full range of responsibilities that came with operating machines used in agricultural production. Typically, *mekhanizator* was expected to attend to machines, do repairs as needed, and work the machines and know the demands and structure of work to be performed. They were mechanics, operators, setters, and tenders all in one. In the specific context of the Soviet agricultural production and female labor, *mekhanizator* routinely referred to tractor drivers who were also in charge of maintaining and repairing tractors as needed.
2 VOANPI, F. 1939. Op. 1. D. 100. L. 455.
3 *Sotsialisticheskoe zemledelie*, 1936, July 17; August 1; *Krest'ianskaia gazeta*, 1937, July 24; *Krest'ianka*, 1982, no. 4, p. 10; Manning, R.T., "Zhenshchiny sovetskoi derevni nakanune Vtoroi mirovoi voiny. 1935–40 gody" in *Otechestvennaia istoriia*, 2001, no. 5, p. 96.
4 *Sotsialisticheskoe zemledelie*, 1936, August 8; Manning, R.T. "Women in the Soviet Countryside on the Eve of World War II. 1935–40" in *Russian Peasant Women*, New York: Oxford University Press, 1992, p. 218.
5 Danilov, V.P., Kim, M.P., Tropkin, N.V. (Eds.), *Sovetskoe krest'ianstvo. Kratkii ocherk (1917–1969)*, M., Izdatel'stvo Politicheskoi literatury, 1970, p. 272.
6 Vyltsan, M.A., *Zavershayushchii etap sozdaniia kolkhoznogo stroia*, M., 1978, p. 114; *Pravda*, 1937, February 12; *Sovetskoe stroitel'stvo*, 1936, no. 1 (January), p. 38; Manning, R.T. "Zhenshchiny sovetskoi derevni nakanune Vtoroi mirovoi voiny. 1935–40 gody," p. 97.
7 Manning, R.T. "Zhenshchiny sovetskoi derevni nakanune Vtoroi mirovoi voiny. 1935–40 gody," p. 97.
8 Slavutskii, A., *Praskov'ia Angelina*, M., 1960, pp. 60–7.
9 *Moskovskii komsomolets*, 2006, September 28.
10 Danilov, V.P., Kim, M.P., Tropkin, N.V. (Eds.), *Sovetskoe krest'ianstvo*, p. 274.

11 Ibid., pp. 314, 318; GARF. F. P-7928. Op. 2. D. 35. L. 93.
12 Danilov, V.P., Kim, M.P., Tropkin, N.V. (Eds.), *Sovetskoe krest'ianstvo*, p. 320.
13 *Izvestiia*, 1943, August 18.
14 *Zhenshchiny Strany Sovetov. Kratkii istoricheskii ocherk*, M., 1977, p. 212.
15 VOANPI, F. 2522. Op. 8. D. 72. L. 2, 6, 13–18.
16 Denisova, L.N., *Zhenshchiny russkikh selenii. Trudovye budni*, M., Mir istorii, 2003, p. 89.
17 *Spravochnik partiinogo rabotnika*, M., 1976, Edition 16, p. 291.
18 *Krest'ianka*, 1980, no. 8, p. 10.
19 *Krest'ianka*, 1982, no. 5, p. 18.
20 *Krest'ianka*, 1977, no. 6, pp. 10–11; 1981, no. 5, pp. 8–9.
21 *Krest'ianka*, 1968, no. 7, pp. 6–7.
22 *Krest'ianka*, 1968, no. 12, pp. 4–5.
23 *Sel'skaia nov'*, 1979, no. 8, p. 7.
24 *Krest'ianka*, 1977, no. 2, p. 2; *Sel'skaia nov'*, 1981, no. 9, p. 19.
25 *SP RSFSR*, 1979, no. 16, Chapter 88.
26 *Resheniia partii i pravitel'stva po khoziaistvennym voprosam*, M., 1972, Tom 8, pp. 119–21.
27 *Sovetskaia pedagogika*, 1979, no. 1, p. 32.
28 Gruzdeva, E.B., Chertikhina, E.S., *Trud i byt sovetskikh zhenshchin*, M., 1983, p. 22.
29 *Krest'ianka*, 1981, no. 10, p. 7; 1987, no. 8, p. 18.
30 *Voprosy epidemiologii i gigieny v Litovskoi SSR: pyti uluchsheniia uslovii truda v sel'skokhoziaistvennom proizvodstve*, Vil'nyus, 1976, p. 190; Fedorova, M., "Ispol'zovani zhenskogo truda v sel'skom khoziaistve" in *Voprosy ekonomiki*, 1975, no. 12, p. 58.
31 *Krest'ianka*, 1987, no. 8, p. 17.
32 Ibid.
33 Ibid., p.18.
34 *Sel'skaia nov'*, 1989, no. 6, p. 7.
35 *Krest'ianka*, 1978, no. 9, p. 4.
36 *Krest'ianka*, 1977, no. 3, p. 21
37 *Krest'ianka*, 1981, no. 5, p. 10.
38 Ibid.
39 Ibid.
40 *Krest'ianka*, 1981, no. 5, p. 9.
41 *Krest'ianka*, 1977, no. 6, p. 11.
42 Most consumer goods that were labeled "deficit" (refrigerators, television sets, winter coats, etc.) were not readily available for sale in stores. Instead, one could only buy these items after a long wait (sometimes years) or in specialized stores that were open only to people with special privileged access to them. Typically these stores served elite members of the Party, although at times access was granted to people who achieved especially high labor output, as well as a number of other categories of people who were deemed worthy.
43 *Krest'ianka*, 1981, no. 5, p. 7; *Sel'skaia nov'*, 1981, no. 9, p. 19; Bridger, S., "Soviet Rural Women: Employment and Family Life" in *Russian Peasant Women*, New York: Oxford University Press, 1992, p. 274.
44 *Sotsial'no-ekonomicheskie problemy agropromyshlennogo kompleksa Rossii*, M., 2000, p. 9.
45 *Golosa krest'ian. Krest'iane Kubani pishut G.V. Iavlinskomu*, Kuban'-M., 1998–99, p. 62; *Sotsiologicheskie obsledovaniia, provedennye v Orlovskoi oblasti v 1993–1996 godakh I. Koznovoi*.

4 Women at the animal wards

1 *Sotsialisticheskoe zemledelie*, 1935, August 16; Manning, R.T., "Women in the Soviet Countryside on the Eve of World War II. 1935–40" in *Russian Peasant Women*, New York: Oxford University Press, 1992, p. 217.
2 Danilov, V.P., Kim, M.P., Tropkin, N.V. (Eds.), *Sovetskoe krest'ianstvo. Kratkii ocherk istorii (1917–1969)*, M., Izdatel'stvo Politicheskaia literatura, 1970, p. 319.
3 *Moskovskii komsomolets*, 2006, September 28.
4 Ibid.
5 RGASPI, F. 591. Op. 1. D. 205. L. 17.
6 VOANPI, F. 2522. Op. 42. D. 74. L. 22; Manning, R.T., "Women in the Soviet Countryside on the Eve of World War II. 1935–40," p. 217.
7 GARF, F. 310. Op. 1. D. 7009. L. 284, 288; D. 7011. L. 116, 120.
8 Manning, R.T., "Women in the Soviet Countryside on the Eve of World War II. 1935–40," p. 217.
9 *Krest'ianka*, 1959, no. 10, p. 18.
10 RGASPI, F. 591. Op. 1. D. 33. L. 303.
11 Staroverov, V.I., *Sotsial'naia struktura sel'skogo naseleniia USSR na etape razvitogo sotsializma*, M., 1978, p. 235; *Sotsial'no-ekonomicheskoe razvitie sela i migratsiia naseleniia*, Novosibirsk, 1972, p. 146; Manning, R.T., "Women in the Soviet Countryside on the Eve of World War II. 1935–40," p. 217; Bridger, S., "Soviet Rural Women: Employment and Family Life" in *Russian Peasant Women*, New York: Oxford University Press, 1992, p. 275.
12 RGASPI, F. 591. Op. 1. D. 34. L. 17.
13 Denisova, L.N., *Zhenshchiny russkikh selenii. Trudovye budni*, M., Mir istorii, 2003, p. 117.
14 GAVO, F. 1300. Op. 1. D. 1463. L. 82, 84–6, 87, 89–91.
15 *Lyudi v gorode i na sele*, M., 1978, p. 64; Bridger, S., "Rural Women and Glasnost" in *Russian Peasant Women*, New York: Oxford University Press, 1992, pp. 296–7.
16 *Sel'skaia nov'*, 1989, no. 7, p. 6.
17 RGANI, F. 5. Op. 24. D. 531. L. 102.
18 *Problemy truda v sel'skom khoziaistve*, M., 1982, p. 43; Mashenkov, V.F., Mal'tsev, I.E., *Formirovanie i ispol'zovanie rabochei sily v sel'skom khoziaistve*, M., 1988, p. 140; *Argumenty i fakty*, 1990, no. 10; *Izvestiia*, 1986, July 4; *Narodnoe khoziaistvo RSFSR v 1989 godu. Statisticheskii ezhegodnik*, M., 1990, p. 561.
19 *Narodnoe khoziaistvo RSFSR v 1989 godu*, p. 561.
20 RGASPI, F. 591. Op. 1. D. 215. L. 79.
21 *Krest'ianka*, 1987, no. 1, p. 17; Bridger, S., *Rural Women and Glasnost*, p. 297.
22 Kharitonova, A., *Svetloe imia—mat'*, M., 1984, pp. 122–3.
23 *Sotsialisticheskoe zemledelie*, 1938, July 24; 1939, February 24; *Kolkhozy vo vtoroi piatiletke*, M., 1938, pp. 60–80; Manning, R.T., "Women in the Soviet Countryside on the Eve of World War II. 1935–40," p. 217; *Komsomol'skaia pravda*, 1988, February 13.
24 *Krest'ianka*, 1980, no. 8, p. 11.
25 *Krest'ianka*, 1987, no. 1, p. 17.
26 Ibid., p. 11; Fedorova, M., "Ispol'zovanie zhenskogo truda v sel'skom khoziaistve" in *Voprosy ekonomiki*, 1975, no. 12, p. 15; Bridger, S., *Soviet Rural Women: Employment and Family Life*, p. 275.
27 Shineleva, L.T., *Zhenshchina i obshchestvo. Deklaratsiia i real'nost'*, M., 1990, p. 41.

28 Kudriashov, V.I., *Organizatsiia truda v molochnom skotovodstve*, M., 1983, p. 114.
29 Tarasevich, V., Leshkevich, V., "Proforientatsiia: opora na sem'yu" in *Molodoi communist*, 1977, no. 7, p. 51.
30 *Krest'ianka*, 1975, no. 11, p. 3.
31 *Krest'ianka*, 1981, no. 10, p. 7.
32 *Trud i upravlenie v sel'skom khoziaistve. Sbornik nauchnykh trudov VNIISXT*, Edition 57, M., 1976, pp. 3–10.
33 Ibid., p. 17.
34 *Gigienicheskie aspekty okhrany okruzhayushchei sredy i zdorov'ia sel'kogo naseleniia v sviazi s intensifikatsiiei sel'skogo khoziaistva*, M., 1977, pp. 32–3, 129–32.
35 Manning, R.T., "Zhenshchiny sovetskoi derevni nakanune Vtoroi mirovoi voiny. 1935–40 gody" in *Otechestvennaia istoriia*, 2001, no. 5, p. 96.
36 "Praskov'ia Evdokiia" in *Krest'ianka*, 1981, no. 10, pp. 2, 8–9.
37 *Krest'ianka*, 1980, no. 12, p. 2.
38 *Argumenty i fakty*, 2004, no. 33.
39 *Argumenty i fakty*, 2004, no. 39.
40 *Argumenty i fakty*, 2004, no. 40; 2004, no. 36.
41 *Argumenty i fakty*, 2004, no. 39.
42 *Rossiiskaia gazeta*, 2009, April 22; *Peredacha NTV May 2 2009 goda; Komsomol'skaia pravda*, 2009, April 29.

5 Women as collective farm leaders and agricultural specialists

1 From numerous speeches by Stalin.
2 VOANPI, F. 1939. Op. 1. D. 198. L. 35 ob.
3 Malukhina, A., "Zhemshchiny v kolkhozakh—bol'shaia sila" in *Sotsialisticheskaia rekonstruktsiia sel'skogo khoziaistva*, 1938, no. 3, March, p. 32; Manning, R.T., *Zhenshchiny sovetskoi derevni nakanune Vtoroi Mirovoi voiny. 1935–1940-e gody*, pp. 93, 101.
4 *Sotsialisticheskoe zemledelie*, 1938, September 9; Manning, R.T., *Women in the Soviet Countryside on the Eve of World War II. 1935–1940*, p. 225.
5 *Pravda*, 1938, February 15; *Sotsialisticheskoe zemledelie*, 1937, October 27, 30; *Partiinoe stroitel'stvo*, 1939, no. 5, March, p. 41; Manning, R.T., *Women in the Soviet Countryside on the Eve of World War II. 1935–1940*, pp. 225–6.
6 Danilov, B.P., Kim, M.P., Tropkin, N.V. (Eds.), *Sovetskoe krest'ianstvo. Kratkii ocherk istorii (1917–1969)*, M., Izdatel'stvo Politicheskoi literatury, 1970, p. 317.
7 Dodge, N.D., Feshbach, M., "The Role of Women in Soviet Agriculture" in *Russian Peasant Women*, New York: Oxford University Press, 1992, p. 250.
8 Khrushchev, N.S., *Sovremennyi etap kommunisticheskogo stroitel'stva i zadachi partii po uluchsheniyu rukovodstva sel'skim khoziaistvom*, M., 1962, p. 259.
9 Bridger, S., "Soviet Rural Women: Employment and Family Life" in *Russian Peasant Women*, New York: Oxford University Press, 1992, p. 280.
10 Kharitonova, A., *Svetloe imia—mat'*, M., 1984, p. 193.
11 *Vestnik statistiki*, 1992, no. 1, p. 64.
12 VOANPI, F. 1939. Op. 1. D. 100. L. 454.
13 VOANPI, F. 1939. Op. 1. D. 160. L. 207.
14 Denisova, L.N., *Zhenshchiny russkikh selenii. Trudovye budni*, M., Mir istorii, 2003, p. 190.
15 Dodge, N.D., Feshbach, M., "The Role of Women in the Soviet Agriculture" pp. 256–7, 270; Danilov, V.P., Kim, M.P., Tropkin, N.V. (Eds.), *Sovetskoe krest'ianstvo. Kratkii ocherk istorii (1917–1969)*, p. 391.
16 RGANI, F. 5. Op. 46. D. 163. L. 15–16.

17 Krapivenskii, S.E., Dement'ev, S.M., Kramarev, V.F., *Sel'skokhoziaist-vennyi kollektiv kak ob'ekt sotsial'nogo planirovaniia*, M., 1981, p. 83.
18 Pankratova, M.G., *Sel'skaia zhenshchina v USSR*, M., 1990, p. 107.
19 *Sel'skaia molodezh'*, 1972, no. 2, p. 17.
20 *Itogi Vsesoyuznoi perepisi naseleniia 1970 goda*, Tom 6, M., 1973, p. 167.
21 *Narodnoe khoziaistvo RSFSR v 1980 godu. Statisticheskii ezhegodnik*, M., 1981, pp. 170, 350; *Narodnoe khoziaistvo RSFSR v 1989 godu. Statisticheskii ezhegodnik*, M., 1990, pp. 276, 590; RGASPI. F. 591. Op. 1. D. 92. L. 29.
22 *Golosa krest'ian. Krest'iane Kubani pishut G.V. Iavlinskomu*, Kuban'-M., 1998–99, p. 5.
23 Riabinina, O., "Muzhik oslab, i vlast' u bab" in *Argumenty i fakty*, 2004, no. 36.
24 *Argumenty i fakty*, 2004, no. 36.
25 *Komsomol'skaia Pravda*, 2009, April 29.

6 Rural intelligentsia

1 *Narodnoe obrazovanie, nauka i kul'tura v SSSR. Statisticheskii sbornik*, M., 1977, p. 9.
2 GARF, F. A-2306. Op. 70. D. 1603. L. 18.
3 *Kul'turnoe stroitel'stvo v SSSR*, M., 1940, pp. 37, 40.
4 GARF, F. P-5462. Op. 18. D. 1603. L. 1, 2; D. 36. L. 8–51.
5 GARF, F. P-5462. Op. 18. D. 1603. L. 1, 2; D. 36. L. 8–51.
6 *Vsesoyuznaia shkol'naia perepis' 15 December 1927 goda*, Tom 1, pp. 24, 240.
7 Kim, M.P., *40 let sovetskoi kul'tury*, M., 1967, p. 154.
8 *Materialy k otchetu obkoma VKP (b) 3–I oblastnoi partkonferentsii Ivanovskoi promyshlennoi oblasti*, M.-Ivanovo, 1932, pp. 156–8.
9 GARF, F. P-8131. Op. 23. D. 202. L. 13–15.
10 Denisova, L.N., *Sud'ba russkoi krest'ianki v XX veke. Brak. Sem'ia. Byt.*, M., Rosspen, 2007, p. 275.
11 Denisova, L.N., *Vseobshchee srednee obrazovanie i sotsial'nyi progress sela*, M., Nauka, 1988, p. 32.
12 Ibid.
13 *Pravda*, 1980, August 12.
14 *Izvestiia*, 1989, July 18.
15 *Sovetskaia pedagogika*, 1979, no. 1, p. 25; GARF. F. 9563. Op. 1. D. 3279. L. 26.
16 GARF, F. 9463. Op. 1. D. 1909. L. 6; D. 1939. L. 32.
17 *Komsomol'skaia Pravda*, 1980, June 8; 1981, February 10, July 15.
18 Denisova, L.N., *Rural Russia: Economic, Social and Moral Crisis*, New York: Nova Science Publishers, 1995, p. 34.
19 *Sotsiologicheskie issledovaniia*, 1991, no. 8, pp. 62–3.
20 *Vsesoyuznaia shkol'naia perepis' 15 December 1927 goda*, Tom 1, pp. 24, 240.
21 *Sel'skaia nov'*, 1989, no. 3, pp. 10–11.
22 *Komsomol'skaia Pravda*, 1989, December 1.
23 Rudenko, I., *V shesti zerkalakh*, M., 1980, p. 60.
24 Ziiatdinova, F.G., "Prestizh professii uchitelia" in *Sotsiologicheskie issledovaniia*, 1991, no. 8, pp. 62–3.
25 *Pravda*, 1990, March 5.
26 Denisova, L.N., *Vseobshchee srednee obrazovanie i sotsial'nyi progress sela*, p. 41.
27 *Prava zhenshchin v Rossii: issledovanie real'noi praktiki ikh soblyudeniia i massovogo soznaniia (po rezul'tatan issledovaniia v g. Rybinske Iaroslavskoi oblasti na osnove glubinnykh interv'yu)*, M., 1998, p. 181.

28 RGAE, F. 4372. Op. 66. D. 2729. L. 4–6; RGASPI. F. 591. Op. 1. D. 160. L. 81–2.
29 *Sel'skaia molodezh'*, 1972, no. 5, p. 21.
30 *Argumenty i fakty*, 2004, no. 30.
31 *Narodnoe obrazovanie, nauka i kul'tura v SSSR*, M., Statisticheskii sbornik, 1977, p. 180.
32 Denisova, L.N., *Rural Russia: Economic, Social and Moral Crisis*, p. 132.
33 *Trud*, 1988, April 6; *Trud*, 1987, May 13; Ibid.; *Sovetskaia kul'tura*, 1989, December 16.
34 Denisova, L.N., *Zhenshchiny russkikh selenii. Trudovye budni*, M., Mir istorii, 2003, p. 214.
35 *Argumenty i fakty*, 2004, no. 45.

7 Migration to cities and the position of newcomers

1 Vishnevskii, A., *Serp i rubl'. Konservativnaia modernizatsiia v USSR*, M., 1998, p. 51.
2 *Sovetskoe krest'ianstvo. Kratkii ocherk istorii (1917–1970)*, M., Izdatel'stvo Politicheskoi literatury, 1973, p. 324; Selunskaia, V.M., *Sotsial'naia struktura sovetskogo obchestva: istoriia i sovremennost'*, M., 1987, p. 82; Arutyunian, U.V., "Kollektivizatsiia sel'skogo khoziaistva i vysvobozhdenie rabochei sily dlia promyshlennosti" in *Formirovanie i razvitie sovetskogo rabochego klassa (1917–1961 gody)*, M., 1964, p. 102; Manning, R.T., "Women in the Soviet Countryside on the Eve of World War II. 1935–40" in *Russian Peasant Women*, New York: Oxford University Press, 1992, p. 210.
3 Danilov, V.P., *Sovetskaia dokolkhoznaia derevnia: naselenie zemlepol'-zovanie, khoziaistvo*, M., 1977, p. 24.
4 RGANI, F. 89. Perechen' 73. doc. 4. L. 3–4.
5 RGANI, F. 89. Perechen' 73. doc. 2. L. 1–2; doc. 107. L. 1.
6 Arkhiv UVD Arkhangel'skoi oblasti. F. 5. Op. 19. d. 861. L. 39.
7 RGANI, F. 89. Perechen' 73. doc. 4. L. 1–2.
8 Zelenin, I.E., "Kul'minatsiia "Bol'shogo terrora" v derevne. Zigzagi agrarnoi politiki (1937–38 gody)" in *Otechestvennaia istoriia*, 2004, no. 1, p. 176.
9 Arkhiv UVD Arkhangel'skoi oblasti. F. 5. Op. 19. D. 857. L. 97.
10 *Rossiisko-britanskoe sotsiologicheskoe obsledovanie rossiiskikh dereven'. 1980–1990-e gody.*
11 Vishnevskii, A., "Na polputi k gorodskomu obchestvu" in *Chelovek*, 1992, no. 1, pp. 14, 17–18.
12 "Ustnye svidetel'stva sel'skoi zhenshchiny Marii Zolotarevoi" in *Demografiia i sotsiologiia*, Edition 15, M., 1996, p. 209.
13 Ibid., pp. 209–11.
14 Malysheva, M., "Kommunisticheskaia utopia v sud'bakh sel'skikh zhenshchin" in *Demografiia i sotsiologiia*, Edition 15, M., 1996, p. 219.
15 Lapidus, G., *Women in Soviet Society*, Berkeley, 1978, p. 166.
16 Popov, V.P., *Ekonomicheskaia politika sovetskogo gosudarstva. 1946–1953 gody*, Tambov, 2000, pp. 21–3, 29–30.
17 Gorbachev, O.V., "Organizovannaia migratsiia iz sela Tsentral'nogo Nechernozem'ia vo vtoroi polovine 1940-kh-1960-e gody" in *Voprosy istorii*, 2003, no. 2, pp. 138–41.
18 Zima, V.F., *Golod v SSSR 1946–1947 godov: proiskhozhdenie i posledstviia*, M., IRI RAN, 1996, pp. 205–7; Popov, V.P., *Ekonomicheskaia politika sovetskogo gosudarstva. 1946–1953 gody*, pp. 17–18, 45, 51.
19 RGANI, F. 5. Op. 15. D. 429. L. 104–9.

20 RGANI, F. 5. Op. 24. D. 530. L. 80–1.
21 Ibid., L. 2–6.
22 RGANI, F. 17. Op. 138. D. 4. L. 62.
23 RGANI, F. 5. Op. 15. D. 426. L. 31.
24 Ibid., L. 92.
25 Ibid., L. 32, 34.
26 RGANI, F. 5. Op. 15. D. 427. L. 15, 16, 18, 25.
27 *Moskoviia. (Prilozhenie k gazete "Moskovskii komsomolets")*, 2004, January 26.
28 RGANI, F. 5. Op. 34. D. 37. L. 73.
29 Ibid., L. 6, 7, 9–11.
30 "Moia zemnaia derevushka. Beseda s V. Astaf'evym" in *Komsomol'skaia Pravda*, 1988, May 12.
31 Butkovskaia, L., "Kak Valia romantiku iskala" in *Krest'ianka*, 1978, no. 7. Oborot oblozhki, p. 1.
32 Dodge, N.D., Feshbach, M., "The Role of Women in Soviet Agriculture" in *Russian Peasant Women,* New York: Oxford University Press, 1992, p. 253.
33 Dodge, N.D., Feshbach, M., op. cit., p. 254.
34 RGASPI, F. 591. Op. 1. D. 164. L. 180.
35 *Vestnik statistiki*, 1991, no. 7, p. 21; 1992, no. 1, p. 15.
36 *Sel'skaia zhizn'*, 1985, August 10; *Sovetskaia Rossiia*, 1985, September 6.
37 *Nash sovremennik*, 1986, no. 3, p. 161; Denisova, L.N., *Rural Russia: Economic, Social and Moral Crisis*, New York: Nova Science Publishers, 1995, p. 172.
38 *Krest'ianka*, 1988, no. 1, p. 12.
39 *Komsomol'skaia Pravda*, 1988, 3, March 8.
40 *Krest'ianka*, 1988, no. 1, p. 12.
41 *Nashi zhenshchiny*, M., 1984, p. 67.
42 *Sotsiologicheskie issledovaniia*, 1987, no. 2, pp. 61–3; *Komsomol'skaia Pravda*, 1988, August 21.
43 Perevedentsev, V.I., *Naselenie SSSR: (vchera, segodnia, zavtra)*, M., 1972, p. 8.
44 *Narodnoe khoziaistvo USSR v 1959 godu. Statisticheskii ezhegodnik*, M., 1960, pp. 28–30; *Narodnoe khoziaistvo SSSR v 1989 godu. Statisticheskii ezhegodnik*, M., 1990, pp. 19–21.
45 *Komsomol'skaia Pravda*, 1988, 3, March 27.
46 *Komsomol'skaia Pravda*, 1988, March 3.
47 *Komsomol'skaia Pravda*, 1988, August 26; *Pravda*, 1989, February 14.
48 VOANPI, F. 2522. Op. 42. D. 74. L. 90–1.
49 RGASPI, F. 591. Op. 1. D. 172. L. 28.
50 RGASPI, F. 17. Op. 102. D. 736. L. 226.
51 *SP SSSR*, 1974, no. 19, Chapter 109.
52 *Vestnik statistiki*, 1984, no. 9, pp. 64, 65.
53 Zotova, O.I., Novikov, V.V., Shorokhova, E.V., *Osobennosti psikhologii krest'ianstva. Proshloe i nastoiashchee*, M., 1983, p. 98.
54 *Izvestiia*, 1964, November 21; *Komsomol'skaia Pravda*, 1975, May 16.
55 *Nash sovremennik*, 1980, no. 3, p. 134.

8 The politics of private life

1 SU RSFSR, 1917, article 11, section 160.
2 *Gendernaia ekspertiza rossiiskogo zakonodatel'stva*, M., 2001, pp. 98–9.
3 SU RSFSR, 1917, article 10, section 152.
4 *Obshchii svod po Imperii rezul'tatov razrabotki dannyh Pervoi vseobshchei perepisi naseleniia, proizvedennoi 28 ianvaria 1897 g*, SPt, 1905, vol. 1, pp. 78, 79.

5 Ekshtut, S., *Na sluzhbe Rossiiskomu Leviafanu. Istoriofilosovskie opyty, glava 8,*
   M., 1998; *Sotsial'naia politika i sotial'naia rabota v izmeniaiushcheisia Rossii,*
   M., 2002, p. 119.
6 Antokol'skaia, M.B., *Semeinoe pravo*, M., 1999, p. 64.
7 SU RSFSR, 1918, article 76, section 318.
8 *Gendernaia ekspertiza rossiiskogo zakonodatel'stva*, M., 2001, p. 100.
9 Genkin, D.M., Novitskii, I.B., Rabinovich, N.V., *Istoriia sovetskogo grazhdanskogo
   prava*, M., 1949, p. 409.
10 *Gendernaia ekspertiza*, p. 102–3.
11 SU RSFSR, 1926, article 82, section 612.
12 SU RSFSR, 1920, article 45, section 205.
13 Antokol'skaia, M.B., op. cit., p. 69.
14 *Gendernaia ekspertiza*, p. 104.
15 SZ SSSR, 1936, article 34, section 309. See also Michaels, Paula, "Motherhood,
   Patriotism, and Ethnicity: Soviet Kazakhstan, and the 1936 Abortion Ban,"
   Feminist Studies, vol. 27, 2001, no. 2, pp. 307–33.
16 Vainshtein, G., "Zabota o zhenshchine-materi i vospitanie detei," *Sovetskoe
   stroitel'stvo,*1936, no. 7, p. 47; Manning, R.T., "Women in the Soviet Countryside
   on the Eve of World War II, 1935–40" in *Russian Peasant Women*, New York:
   Oxford University Press, 1992, pp. 208–9.
17 *Pravda*, 1936, December 21; 1937, February 18 and March 3; and 1938, April 8.
18 Ransel, D.L., *Village Mothers: Three Generations of Change in Russia and
   Tataria*, Bloomington, IN: Indiana University Press, 2000.
19 *Sotsialisticheskoe zemledelie*, 1937, June 24.
20 *Krestianskaia gazeta*, 1936, June 20 and 22; Manning, R.T., "Women in the Soviet
   Countryside on the Eve of World War II, 1935–1940," pp. 209–10.
21 Antokol'skaia, M.B., *Semeinoe pravo*, pp. 72–3.
22 *Sovetskoe krestianstvo. Kratkii ocherk istorii (1917–70)*, M., 1973, p. 324;
   Selunskaia, V.M., *Sotsial'naia struktura sovetskogo obshchestva: istoriia i
   sovremennost'*, M., 1987, p. 82; Arutiunian, Iu.V., "Kollektivizatsiia sel'skogo
   khoziaistva i vysvobozhdenie rabochei sily dlia promyshlennosti" in *Formirovanie
   i razvitie sovetskogo rabochego klassa (1917–1961gg)*, M., 1964, p. 102.
23 *Pravda*, 1936, June 9; *Krestianskaia gazeta*, 1936, June 8; Manning, R.T., "Women
   in the Soviet Countryside on the Eve of World War II, 1935–1940," p. 211.
24 Ibid.
25 *Vedomosti verkhovnogo soveta SSSR*, 1957, no. 37.
26 Ibid., article 17, section 447.
27 GARF, F. R-8131, Op. 22, D. 34, L, 7, 18.
28 GARF, F. R-8131, Op. 22, D. 33, L. 9.
29 *Vedomosti verkhovnogo soveta SSSR*, 1945, no. 15
30 Sudebnaia praktika, 1951, article 1, sections 7–8.
31 *Biulluten' verkhovnogo suda SSSR*, 1961, 31, p. 7–9.
32 Kavamoto, K., *Dvadtsatiletnaia popytka izmenit' pravovoe polozhenie detei,
   rodivshikhsia ne v zaregistrirovannom brake posle Velikoi Otechestvennoi voiny,*
   M., 2002, pp. 14–15.
33 *Sovetskaia zhizn', 1945–1953*, M., 2003, p. 680.
34 Antipova, M.L., "Vnebrachnye deti v Rossii" in *Statisticheskoe issledovanie
   sotial'no-demograficheskoi situatsii. Sbornik nauchnykh trudov*, M., 2002, p. 5.
35 Verbitskaia, O.M., *Naselenie rossiiskoi derevni v 1939–1959 gg. Problemy
   demograficheskogo razvitiia*, M., 2002, pp. 147–8.
36 Antakol'skaia, M.B., op. cit., pp. 75–6.
37 GARF, F. R-8131, Op. 23, D. 239, L. 2; Op. 22, D. 34, L. 48.
38 GARF, F. R-8131, Op. 22, D. 34, L. 51–3, 56, 58–9.

39 *Vedomosti verkhovnogo soveta SSSR*, 1947, no. 10.
40 The Soviet government consistently failed to provide its citizens with effective contraceptives even much later in the twentieth century. As a result, abortions to the last day of the Soviet Union were perceived as contraceptive measures, and it was common to come across women who had six or more abortions in their lifetime.
41 Kavamoto, K., *Dvadtsatiletnaia popytka izmenit' pravovoe polozhenie detei, rodivshikhsia ne v zaregistrirovannom brake posle Velikoi Otechestvennoi voiny*, pp. 15–17.
42 Ibid.
43 *Izvestiia Narkomata truda SSSR*, 1930, no. 10, p. 234.
44 SP SSSR, 1940, no. 19. p. 465; 1941, no. 15, p. 282.
45 *Vestnik statistiki*, 1991, no. 8, pp. 53, 54.
46 RGASPI, F. 591, Op. 1, D. 96, L. 38.
47 Denisova, L.N., *Ischezaiushchaia derevnia Rossii*, pp. 184–5.
48 *Reshenie partii i pravitel'stva po sel'skomu khoziaistvu (1960–1974gg)*, M., 1975, p. 662.
49 *Zakonodatel'stvo o pravakh zhenshchin v SSSR*, pp. 36–7.
50 *Vedomosti verkhovnogo soveta SSSR*, 1969, no. 32, section 1,086.
51 *Konstitutsiia Soiuza Sovestkikh Sotsialisticheskikh Respublik*, M., 1977.
52 Ibid.
53 Polenina, S.V., "Polozhenie zhenshchiny v obshchestve i semie," *Zhenshchiny v sovremennom mire*, M., 1989, p. 65.
54 *Trud, semia, byt sovetskoi zhenshchiny*, M., 1990, pp. 11–12.
55 Tolkunova, V.N., "Sovetskoe zakonodatel'stvo v period 10-letiia zhenshchiny OON" in *Rol' zhenshchiny v sovremennom obshchestve. K itogam 10-letiia OON. Sbornik statei*, M., 1985, p. 26.
56 *Vedomosti verkhovnogo soveta SSSR*, 1982, no. 25, section 464.
57 *Sotsial'naia politika i sotsial'naia rabota v izmeniaiushcheisia Rossii*, M., 2002, pp. 157–8.
58 "Postanovlenie Verkhovnogo Soveta RSFSR 'o neotlozhnykh merakh po uluchsheniiu polozheniia zhenshchin, semie, okhrany materinstva i detstva na sele' ot 1 noiabria 1990g.," *Polozhenie Soveta Narodnykh deputatov I Verkhvnogo Soveta RSFSR*, 1990, article 24, section 287.
59 "Postanovlenie Verkhovnogo Soveta RSFSR ot 25 ianvaria 1991 g 'o poriadke primenenia postanovleniia o neotlozhnykh merakh po uluchsheniiu polozheniia zhenshchin, semie, okhrany materinstva i detstva na sele' ot 1 noiabria 1990g.," *Polozhenie Soveta Narodnykh deputatov I Verkhvnogo Soveta RSFSR*, 1991, article 6, section 89.
60 *Semeinyi Kodeks Rossiiskoi Federatsii*, M., 1996.

9 Marriages

1 Chudnovskikh, S.L., "Ocherki narodnogo iuridicheskogo byta Altaiskogo gornogo okruga,' *Russkoe bogatstvo*, 1984, no. 7, p. 34.
2 Neklepaev, I.Ia., "Poveria I obychei Surgutskogo kraia," *Zapiski Zap-Sib.Otdela RGO*, Omsk, 1903, vol. 30, pp. 140–4.
3 Minenko, N.A., Rabtsevich, V.V., *Liubov I semmia u krestian v starinu: Ural I Sibir' v XVIII–XIX vekakh*, Cheliabinsk, 1997, p. 51.
4 Krivoshapkin, M.F., *Eniseiskii okrug i ego zhizn*. SPt, 1865, vol. 1, pp. 54–5.
5 Minenko, Rabtsevich, *Liubov I semmia u krestian v starinu*, p. 283.
6 *Golosa krestian: Sel'skaia Rossiia XX veka v krestianskikh memuarakh*, M., 1996, p. 241.

7 Lipinskaia, V.A., Safianova, A.V., "Svadebnye obriady russkogo naseleniia Altaiskogo gornogo okruga," *Russkii narodnyi svadebnyi obriad*, Leningrad, 1978, p. 185.
8 *Golosa krestian*, pp. 80–1.
9 Belov, V., *Medovyi mesiats*, M., 2002.
10 Anokhina, L.A. *Kul'tura i byt kolkhoznikov Kalininskoi oblasti*, p. 225.
11 Rossiisko-britanskoe sotsiologicheskoe obsledovanie rossiiskikh dereven' 1980–90-kh gg.
12 Ibid.
13 *Selo Viriatino v proshlom i nastoiashchem. Opyt etnograficheskogo izucheniia russkoi kolkhoznoi zhizni*, M., 1958, p. 226.
14 Kharitonova, A., *Svetloe imia—mat'*, M., 1984, p. 6.
15 Verbitskaia, O.M., *Naselenie rossiiskoi derevni v 1939–1959gg. Problemy demograficheskogo razvitiia*, pp. 100–3.
16 *Selo Vipiatino*, pp. 226–7.
17 *Krestianka*, 1981, no. 10, pp. 26–7.
18 Bridget, Susan, *Soviet Rural Women: Employment and Family Life*, p. 281.
19 Vlasova, I.V., Shmeleva, M.N., "Semeino-brachnye otnoshenia v 1917–90-h godakh" in *Russkie*, M., 1997, p. 439.
20 Fainburg, Z.I., "K voprosu ob eticheskoi motivatsii braka" in *Sotsial'nye issledovaniia*. M., 1970, pp. 68, 70.
21 Pankratova, M.G., *Sel'skaia zhenshchina v SSSR*, M., 1990, pp. 123–5.
22 Ibid.
23 Rossiisko-britanskoe sotsiologicheskoe obsledovanie rossiiskikh dereven' 1980–90-kh gg.
24 Kharchev, A.G., *Brak i semia v SSSR. opyt sotsiologicheskogo issledovaniia*, M., 1964, p. 215; Chuiko, L.B., *Braki i razvody*, M., 1975, p. 98.
25 Pankratova, M.G., *Sel'skaia zhenshchina v SSSR*, pp. 123–5.
26 *Selo Viriatino v proshlom i nastoiashchem*, p. 226.
27 Ibid.
28 Pankratova, M.G., *Sel'skaia zhenshchina v SSSR*, p. 120–3.
29 Verbitskaia, O.M., *Naselenie rossiiskoi derevni v 1939–1959gg. Problemy demograficheskogo razvitiia*, p. 65.
30 Soloviev, N.A., *Semia v sotsialisticheskom obshchestve*, M., 1981, pp. 33–5.
31 Pankratova, M.G., *Sel'skaia zhenshchina v SSSR*, pp. 71–2.

10 Conflicts and divorces

1 Verbitskaia, O.M., *Naselenie rossiiskoi derevni v 1939–1959gg. Problemy demograficheskogo razvitiia*, M., 2002, pp. 228–9.
2 Vlasova, I.V., Shmeleva, M.N., "Semeino-brachnye otnoshenia v 1917–1990-h godakh," p. 439.
3 Verbitskaia, O.M., *Naselenie rossiiskoi derevni v 1939–1959gg. Problemy demograficheskogo razvitiia*, pp. 230–1; *Riazanskoe selo Korablino (Istorria, ekonomika, byt, kul'tura, liudi sela)*, Riazan', 1957, p. 144.
4 Volkov, A.G., "Semia kak factor izmeneniia demograficheskoi situatsii," *Sotsiologicheskie issledovania*, 1981, no. 1, p. 38.
5 *Vestnik statistiki*, 1991, no. 8., pp. 52, 64; *Chelovek i trud*, 1992, no. 3, p. 11.
6 *Sotsial'no-ekonomicheskie problem agropromyshlennogo kompleksa Rossii*, M., 2002, p. 21.
7 *Naselenie Rossii 1996. Ezhegodnyi demograficheskii doklad*, M., 1997, p. 68.
8 Burova, N.S., *Sotsiologiia i pravo o razvode*, Minsk, 1979, p. 29.
9 Golod, S.I., *Semia i brak: Istoriko-sotsiologicheskii analyz*, SPt, 1998, p. 78.

10 Sysenko, V.A., *Ustoichivost' braka. Problemy, factory, usloviia*, M., 1981, pp. 140–1.
11 Chuiko, L.B., *Braki i razvody*, M., 1975, p. 139; Sedel'nikov., S.S., "Pozitsii suprugov i tipologicheskie osobennosti reaktsii na razvod," *Sotsialogicheskie issledovaniia*, 1992, no. 2, p. 40.
12 *Sotsiologicheskie issledovaniia*, 1976, no. 3, p. 78; *Kollektiv kolkhoznikov*, M., 1970, pp. 215–16.
13 RGANI, F. 5, Op. 33, D. 217, L. 1.
14 *Sel'skaia molodezh'*, 1987, no. 12, p. 23.
15 RGANI, F. 5, Op. 15, D. 690, L. 78.
16 RGANI, F. 5, Op. 33, D. 66, L. 92.
17 Ibid., L. 90.
18 RGANI, F. 5, Op. 33, D. 217, L. 1–2.
19 Ibid., L. 5.
20 Korolev, Iu.A., *Brak i razvod (sovremenniy tendentsii)*, M., 1978, p. 131.
21 *Krestianka*, 1981, no. 7, p. 27.
22 *Sotsialogicheskie issledovaniia*, 1986, no. 6.
23 Bridget, Susan, *Soviet Rural Women: Employment and Family Life*, p. 286.
24 Rossiisko-britanskoe sotsiologicheskoe obsledovanie rossiiskikh dereven' 1980–90-kh gg. Istoriia semie Red'kinykh iz derevni Leont'evshchina Kichmengsko-Gorodetskogo raiona Volorodskoi oblasti.
25 Ibid.
26 *Russkie. Etnosotsiologicheskie ocherki*, M., 1992, pp. 164–5.
27 *Krestianka*, 1982, no. 10, pp. 26–7.
28 *Krestianka*, 1982, no. 1, p. 29.
29 *Krestianka*, 1982, no. 1, p. 30.
30 *Krestianka*, 1989, no. 11, pp. 36–7.
31 *Krestianka*, 1982, no. 1, p. 29.
32 E.g., *Krestianka*, 1980, no. 4, p. 26.
33 VOANPI, F. 2522, Op. 17, D. 85, L. 27–9.

11 Domostroi

1 A Russian proverb.
2 Gurevich, I., Pavlovich, B., *Istoricheskaia khrestomatiia po russkoi istorii* (in 3 volumes), SPt.: Tip. Ministersva putei soobshcheniia, 1989, Ch 2, pp. 53–5.
3 Semenova Tyn'-Shanskaya, O.P., *Zhizn' "Ivana." Ocherki iz byta krest ian odnoi iz chernozemnykh gubernii*, SPt, 1914, p. 5.
4 Zemtsov, L.I., "Krest'ianki v volostnom sude" in *Sotsial'naia istoriia rossiiskoi provintsii v kontekste modernizatsii agrarnogo obchestva v XVIII–XX vekakh. Materialy mezhdunarodnoi konferentsii*, Tambov, 2002, p. 328.
5 Berezanskii, P., *Obychnoe ugolovnoe pravo krest'ian Tambovskoi gubernii*, Kiev: Universitetskaya tip., 1880, pp. 217–18.
6 Bezgin, V.B., *Krest'ianskaya povsednevnost' (Traditsii kontsa XIX-nachala XX veka)*, M.-Tambov: Izdatel'stvo TGTU, 2004, p. 222.
7 Zheleznov, F., *Voronezhskaya derevnya. Bol'she-Vereiskaia volost'*, Edition 2, Voronezh, 1926, p. 28.
8 Zhiromskaya, V.B., "Otnohenie naseleniia k religii: po materialam perepisi 1937 goda" in *Trudy Instituta Rossiiskoi istorii RAN. 1997–1998 gody*, Edition 2, M.: ROSSPEN, 2000, p. 329.
9 *Khudozhestvennyi kinofil'm "Chlen pravitel'stva,"* Lenfilm, 1939.
10 *Vestnik statistiki*, 1991, no. 8, pp. 65–6
11 VOANPI, F. 2522, Op. 63, D. 51, L. 211–13.

12 "Kontseptsiya ulucheniya polozheniya zhenhin v Rossiiskoi Federatsii," *Rossiiskaya gazeta*, 1996, February 14.
13 Heise, L.L., Pitanguy, J., Germain, A., "Violence against women. The hidden health burden," *World Bank Discussion Papers*. The World Bank. Washington, DC, 1994.
14 Ibid. Also Gelles R.J., Harrop, J.W., "Violence, battering and psychological distress among women," *Journal of Interpersonal Violence*, 1989, no. 4, pp. 400–20; Mullen, P.E., Romans-Clarcson, S.E., Waltson V.A., Herbison, P.E., "Impact of sexual and physical abuse on women's mental health," *Lancet*, 1988, pp. 841–5; Ratner, P.A., "The incidence of wife abuse and mental health status in abused wives in Edmonton, Alberta," *Canadian Journal of Public Health*, 1993, pp. 84, 246–9.
15 Stark E., Flitcraf, A., "Violence against intimates: an epidemiological review" in Van Hasselt, V.B., Morrison, R.L., Bellack, A.S., Hersen, M. (Eds.), *Handbook of Family Violence*, New York, 1988, pp. 293–318.
16 Arbatova, M., "'Otkrytoe pis'mo' pri uchastii v predvybornoi kampanii v Gosudarstvennuiy dumu," *Proshchanie s XX vekom*. Book 2, M.: EKSMO, 2002, p. 323.
17 Nikolaeva, Y., "Romanticheskii sadism," *Argumenty i fakty*, 2000, no. 37, p. 17.
18 Ibid.
19 Metelitsa, G., "V derevnykh ludi terpyat bol' do poslednego. Sel'skii dokhtur," *Argumenty i fakty*, 2004, no. 30.
20 *Argumenty i fakty*, 2000, no. 37.
21 *Zhenshchina novoi Rossii: Kakaya ona? Kak zhivet? K chemu stremitsia?* M., 2002, pp. 78–83, 115.
22 Titov, U.P., *Khrestomatiya po istorii gosudarstva i prava Rossii. Uchebnoe posobie*, M., 1997, p. 193.
23 *Giberniia* is an administrative territorial unit in the Russian Empire, 1764–1929.
24 Bezgin, V.B., *Krest'yanskaya povsednevnost' (Traditsii kontsa XIX–nachala XX veka)*, M-Tambov: Izdatel stvo Tambovskogo gosudarstvennogo universiteta, 2004, p. 295.
25 *Vedomosti Verkhovnogo Soveta SSSR*, 1962, no. 8, article 84.
26 Geztsenzon, A.A., *Ugolovnoe pravo i sotsiologiya*, M.: Uridicheskaya literatura, 1970, p. 79.
27 *Gendernaya ekspertiza rossiiskogo zakonodatel'stva*, M.: Izdatel'stvo BEK, 2001, p. 192.
28 Article 131 of the Criminal Code.
29 Luneev, V.V., (Zaveduiushchii sektorom ugolovnogo prava i kriminologii Instituta gosudarstva i prava Rossiiskoi Akademii Nauk), *Prestupnost' XX veka. Mirovye, regional' nye i rossiiskie tendentsii. Mirovoi kriminologicheskii analiz*, M., 1997, pp. 216, 218; GARF, F. 949, Op. 3, D. 171, L. 63–4.
30 "Kontseptsiia uluchsheniia polozheniia zhenshchin v Rossiiskoi Federatsii," *Rossiiskaya gazeta*, 1996, February 14; *Kriminologiya*, M., 1999, p. 424.
31 *Prestupnost i reformy v Rossii*, p. 161.
32 Luneev, V.V. *Prestupnost' XX veka*, p. 136; Hur, E.M., *Nashe prestupnoe obshchestvo*, M., 1977, pp. 267–8.
33 Koss, M., "The Women's health research agenda. Violence against women," *American Psychologist*, 1990, no. 3, pp. 374–80; Heise, L.L, Raikes, A., Watts, C.N., Zwi, A.B., "Violence against women: A Neglected public health issue in less developed countries," *Social Science and Medicine*, 1994, no. 39 (9), pp. 1165–79.
34 Luneev, V.V., *Prestupnost XX veka*, p. 210.

35 Bezgin, V.B., *Krest'ianskaya povsednevnost (Traditsii kontsa XIX-nachala XX veka)*, M-Tambov: Izdatel'stvo Tambovskogo gosudarstvennogo universiteta, 2004, p. 295.
36 Zdravomyslova, E.A., Temkina, A.A., *Transformatsiya gendernogo grazhdanstva v sovremennoi Rossii, Kuda prishla Rossiya?Itogi sotsiental'noi transformatsii. Mezhdunarodnyi simpozium 16–18 yanvarya 2003 goda*, M., 2003, pp. 148–9.

12 Alcoholism in the countryside

1 Ozerov, H.S., *Alkogolizm i bor'ba s nim*, M., 1914, p. 85.
2 Astyrov, H.M., *V volostnykh pisaryakh. Ocherki krest' yanskogo samoupravleniya*, M., 1898, p. 268.
3 Ptitsin, V., *Obychnoe sudoproizvodstvo krest'yan Saratovskoi gubernii*, SPt, 1886, p. 86.
4 *Rossiisko-britanskoye sociologicheskoye issledovanie dereven', 1980–1990-e gody.*
5 Gromyko, M.M., *Mir russkoy derevni*, M.: Molodaya gvardiya, 1991, p. 114.
6 Takala, I.R., *"Veselie Rusi." Istoriya alkogolnoi problemy v Rossii*, SPt.: Zhurnal "Neva," 2002, p. 170.
7 See the discussion of migration for more information.
8 RGASPI, F. 591, Op. 1, D. 225, L. 43.
9 RGASPI, F. 591, Op. 1, D. 91, L. 18–19.
10 RGASPI, F. 591, Op. 1, D. 225, L. 44.
11 RGASPI, F. 591, Op. 1, D. 59, L. 54–5.
12 RGANI, F. 5, Op. 33, D. 217, L. 12.
13 RGANI, F. 5, Op. 33, D. 233, L. 79–80.
14 RGANI, F. 5, Op. 33, D. 233, L. 124–30.
15 RGANI, F. 5, Op. 33, D. 233, L. 63.
16 *Krest'ianka*, 1979, no. 12, p. 20.
17 Vinogradskii, V., Vinogradskaya, O., Nikulina, A., Fadeeva, O., "Istoriya sel skoi zhenshchiny: sem'ia, khoziaistvo, budzhet," *Refleksivnoe krest'ianovedenie. Desyatiletie issledovanii selskoi Rossii*, M.: MVSHSEN, ROSSPEN, 2002, p. 235.
18 RGASPI, F. 591, Op. 1, D. 244, L. 38.
19 Ibid., L. 97.
20 RGANI, F. 5. Op. 33, D. 217, L. 20.
21 RGASPI, F. 591, Op. 1, D. 38, L. 61–2; RGASPI. F. 591, Op. 1, D. 244, L. 62.
22 Ryzhkov, N.I., *Desyat' let velikikh potryasenii*. M.: Associaciya "Kniga, prosvechenye, miloserdiye," 1995, p. 93.
23 Gorbachev, M.S., *Zhizn' i reformy*, Book 1, M.: Novosty, 1995, p. 340.
24 GARF, F 5446, Op. 147, D. 374, L. 32.
25 Ryzhkov, N.I., *Desyat' let velikikh potryasenii*, p. 101.
26 Krikov, M., "Nostal'giia po chasu volka," *VTSIOM*, 2005, May 20.
27 *Krest'ianka*, 1979, no. 10, p. 27.
28 For examples, see RGASPI, F. 591, Op. 1, D. 39, L. 117.
29 Taskala, I.R., *"Veselie Rusi." Istoriya alkogolnoi problemy v Rossii*, p. 254.
30 RGASPI, F. 591, Op. 1, D. 91, L. 22–3, 26–7.
31 RGASPI, F. 8, List 12, D. 6, L. 4.
32 Koss, M., "The Women's health research agenda. Violence against Women," *American Psychologist*, 1990, no. 3, pp. 374–80; Heise, L.L., Raikes, A., Watts, C.N., Zwi, A.B., "Violence against women: A neglected public health issue in less developed countries," *Social Science and Medicine*, 1994, no. 39(9), pp. 1165–79.
33 RGASPI, F. 591, Op. 1, D. 91, L. 19–20.
34 *Narodnoe khozyastvo RSFSR v 1989 godakh*, pp. 310, 314; GARF, F. 3492, Op. 3, D. 171, L. 63–4.

35 *Zhenshchiny v USSR*, 1990, p. 18.
36 Gorbachev, M.S., "O zadachakh po korennoi perestroikae upravleniya economikoi," *Pravda*, 1987, June 26; Levin, B.M., *Mnimye potrebnosti*, M.: Politizdat, 1986, p. 18; Zaigraev, G.G., *Obch'estvo i alkogol'*, M.: NII MVD RF, 1992, p. 43; Takala, I.R., *"Veselie Rusi." Istoriya alkogolnoi problemy v Rossii*, p. 258; Grigor'eva, N.S., Chubarova, T.V., *Gendernyi podkhod v zdravookhranenii*, M.: Al'fa-print, 2001, p. 56.
37 Kur'ianova, N.N., Serdukov, A.G., *Mediko-sotsial'nye aspekty zhenskogo alkogolizma*, Astrakhan': Izdatel'stvo AGMA, 2002, p. 5.
38 Kur'ianova, N.N., Serdukov, A.G., *Mediko-sotsial'nye aspekty zhenskogo alkogolizma*, p. 31.
39 Kur'ianova, N.N., Serdukov, A.G., *Metodiko-sotsial'nye aspekty zhenskogo alkogolizma*, pp. 5, 29–33, 67–9, 89–94; Grigor'eva, N.S., Chubarova, T.V., *Gendernyi podkhod v zdravookhranenii*.
40 *Zhenshchiny v SSSR*, 1990, p. 18.
41 Potapov, V., "Neuteshitel'naia situatsiia (Problemy alkogolizma i narkomanii v Rossii)," *Trezvost 'i kul'tura*, 2002, no. 3, p. 7.
42 Takala, I.R., *"Veselie Rusi." Istoriya alkogolnoi problemy v Rossii*, pp. 282–3.
43 *Golosa krest'ian. Krest'iane Kubani pishut G. Yavlinskomu*, Kuban'-M., 1998–99, p. 3.
44 *Peredacha rossiiskogo televideniya "Moment istiny,"* 2006, September.
45 Kriukov, M. "Nostal'giia po chasu volka," *VTSIOM*, 2005, May 20.

13 The female face of the criminal world

1 Ivnitskii, N.A., "Golod 1932–33 godov: Kto vinovat?" in *Golod 1932–1933 godov*, M.: Gos. Komitet RF po vysshemu obrazovaniy, 1995, p. 64; "M.A. Sholokhov—I.V. Stalinu 4 aprelya 1933 goda," *Sud'by rossiiskogo krest'ianstva*, M., 1996, p. 550.
2 Ivnitskii, N.A., "Kollectivizatsiya i rasculachivanie v nachale 30-kh godov," *Kooperativnyi plan. Illuzii i real'nost'*, M., PGGU, 1995, pp. 63–4.
3 *Tragediya sovetskoi derevni. Kollectivizatsiya i rasculachivanie. Dokumenty i materially*. V 5 Tomakh. Tom 5. Book 1, 1937, M: Rosspen, 2004, pp. 244, 369, 387, 393; Book 2, Doc. no. 17, 66, 144, 145.
4 Zemskov, V.N., *Spetsposelentsy v SSSR. 1930–1960*, M., Nauka, 2003.
5 Mukhina, I., *Germans of the Soviet Union*, London and New York: Routledge, 2007.
6 *Zipun* is the main type of male upper garment dating back to medieval Russia, typically worn over a long shirt. *Zipun* was often cited to emphasize that peasants', and especially kulaks', lifestyle was still rooted in "ancient" and "backward" tsarist Russian traditions.
7 Ivnitskii, N.A., *Kollectivizatsiya i rasculachivanie v nachale 30-kh godov*, p. 61.
8 For further information, refer to Viola, L., *The Unknown Gulag: The Lost World of Stalin's Special Settlements*, Oxford University Press, 2007; Mukhina, I., *Germans of the Soviet Union*, Routledge, 2007.
9 Ivanova, G.M., *Gulag v sisteme totalitarnogo gosudarstva*, M.: ZAO "Pervyi pechatnyi dvor," 1997, p. 111.
10 Ivanova, G.M., *Istoriia Gulaga. 1918–1958. Sotsial'no-ekonomicheskii i politiko-pravovoi aspekty*, M.: Nauka, 2006, p. 281.
11 Zima, V.F., *Golod v USSR 1946–1947 godov: proiskhozhdenie i posledstviya*, M. IRI RAN, 1996, pp. 116, 128.
12 Ivanova, G.M. *Istoriia Gulaga. 1918–1958*, p. 282.
13 Zima, V.F., *Golod v USSR 1946–1947 godov: proiskhozhdenie i posledstviya*, pp. 119, 128.

14 Ivanova, G.M., "Zhenshchiny v zakluchenii (istoriko-pravovoi aspect)" in *Zhenshchina. Gender. Kul'tura*, M.: RLSHGI, 1999, p. 278.
15 Luneev, V.V., *Prestupnost' XX veka. Mirvoi kriminologicheskii analiz*, pp. 13, 35.
16 Ibid., pp. 21, 26.
17 Ibid., pp. 34–5.
18 Shabanova, M.A., "Individual'naia i sotsial'naia svobody v reformiruemoi Rossii" in *Kuda idet Rossiia?Vlast', obshchestvo, lichnost'. Mezhdunarodnyi simpozium 17–18 yanvarya 2000 goda*, M., 2000, pp. 406–7.
19 Luneev, V.V. *Osobennosti sovremennoi prestupnosti v Rossii*, pp. 266–8.
20 *Vestnik statistiki*, 1991, no. 8, pp. 64–7.
21 Shabanova, M.A., "Individual'naia i sotsial'naia svobody v reformiruemoi Rossii," p. 407.
22 *Sotsial'no-economicheskie problemy agropromyshlennogo kompleksa Rossii*, M., 2000, p. 21.
23 *Golosa krest'ian. Krest'iane Kubani pishut G. Yavlinskomu*, Kuban'-M., 1998–99, p. 6.
24 *Gendernaya ekspertiza rossiiskogo zakonodatel'stva*, pp. 214–19.
25 Ivanova, G.M. *Zhenshchiny v zakluchenii (istoriko-pravovoi aspect)*, p. 281.
26 *Gendernaya ekspertiza rossiiskogo zakonodatel'stva*, p. 229.
27 *Rossiiskii statisticheskii ezhegodnik. 2002*. Statisticheskii sbornik, M., Goskomstat Rossii, 2002, pp. 273–5.

14 Women of the oldest profession

1 *Byt velikorusskikh krest'ian-zemledel'tsev*. Opisanie materialov etnograficheskogo buro kn. V.N. Tenisheva, M., 1993, p. 276.
2 D'iachenko, A.P., "Interdevochki v zerkale pressy," *Prostitutsiya i prestupnost'. Problemy. Diskussii. Predlozheniya*, M., 1991, pp. 71–9; *Zhenshchiny v SSSR*, 1990, p. 18.
3 Rak, L., "V nochnom tupike," *Studencheskii meridian*, 1988, no. 9.
4 Gilinskii, Y.I., "Prostitutsiya kak ona est," *Prostitutsiya i prestupnost'*, pp. 99–122; Maidanskaya, N., "Proza 'sladkoi zhizni'," *Sovetskaya kul'tura*, 1990, March 24.
5 *Kriminologiia*, M., 1999, p. 492.
6 Gilinskii, pp. 99–122.
7 "Monolog nochnoi fei," *Trezvost' i kul'tura*, 1988, no. 9.
8 Kislinskaya, L., "'Legkoe povedenie' na vesakh pravosudiya," *Sovetskaya Rossiia*, 1987, March 12; Alekseev, B., "Nochnoi promysel," *Nedelia*, 1987, no. 12; Lilayants, I., "'Dollar girl'. Novaya moskovskaya elita," *Shchit i mech*, 1991, no. 52, December 26.
9 Aivazova, S., Kertman, G., *Zhenshchiny na randevu s rossiiskoi democratici*, M.: ESLAN, 2001, p. 14.
10 *Sotsiologicheskie issledovaniya*, 1992, no. 5, p. 116.
11 Turukanova, E., "Zhenshchiny v poiskakh raboty za rubezhom: zhenskaya trudovaya migratsiya iz Rossii" in *Valdai-96. Materialy pervoi letnei shkoly zhenskig i gendernykh issledovanii*, M.: MTSGI, 1997.
12 *Prestuplenie i porabochenie. Razoblachenie seks-torgovli zhenshchinami iz stran byvshego USSR*, p. 5.

15 Religion

1 For the remaining percentage, 10 percent were Old Believers; 8 percent were Catholics; 4.5 percent Protestants; 6 percent Muslim; and 4 percent Jews.
2 Vasil'eva, O.U., *Russkaya Pravoslavnaya Tserkov' v politike Sovetskogo gosudarstva v 1943–1948 godakh*, M.: IRI RAN, 2001, pp. 9, 13; Vasil'eva, O.U.,

Istoriya Pravoslavnoi tserkvi na stranitsakh shkol'nogo ychebnika, "Kruglyi stol": Kakim byt' sovremennomu shkol'nomu uchebniku po Otechestvennoi istorii XX veka? *Otechestvennaya istoriya*, 2002, no. 3, p. 45.

3   This number includes all believers, not just Russian Orthodox.

4   Zhiromskaya, V.B., "Otnoshenie naseleniya k religii: po materialam perepisi 1937 goda," *Trudy Instituta Rossiiskoi istorii RAN, 1997–1998 gody*, Edition 2, M., 2000, pp. 329–30. In the Soviet census of 1937, 80 percent of all respondents aged 16 and older listed their religious affiliation. (Zhiromskaya, V.B., "Otnoshenie naseleniya k religii: po materialam perepisi 1937 goda," p. 325.)

5   Zhiromskaya, V.B., *Demograficheskaya istoriya Rossii v 1930-e gody. Vzglyd v neizvestnoe*, M.: ROSSPEN, 2001, pp. 206, 209.

6   Voroshilova, O., "Permskii pravoslavnyi prikhod v epokhu peremen (1840-e–1940-e gody)," *Vpered* (g. Ustyzhna Vologodskoi oblasti), 2002, May 31.

7   Chuikova, T.B., "Likvidatsiya tserkvei v Borovichskom raione v 1935–42 godakh. Po dokumentam Gosudarstvennogo arkhiva Novgorodskoi oblasti," *Regional'nye aspekty istoricheskogo puti pravoslaviya: arkhivy, istochniki, metodologiya issledovanii. Istoricheskoe kraevedenie i arkhivy*, Edition 2, Vologda, 2001, pp. 469–73.

8   Arkhiv UVD Arkhangel'skoi oblasti, F. 5, Op. 19, D. 857, L. 11, 28, 31.

9   Beglov, A.L., *Tserkovnoe podpol'e 1920–1940-kh godov v USSR v kontekste gosudarstvenno-tserkovnykh otnoshenii. Dissertatsiya na soiskanie kandidata istoricheskikh nauk*, M., 2003.

10   Vasil'eva, O.U. *Russkaya Pravoslavnaya Tserkov' v politike Sovetskogo gosudarstva v 1943–1948 godakh*, pp. 33–4.

11   RGANI, F. 89, Perechen' 49, D. 24, L. 1–3.

12   RGANI, F. 89, Perechen' 49, D. 25, L. 2–4.

13   Ibid., L. 2.

14   Otechestvennaya istoriya, 2002, no. 3, p. 45.

15   Vasil'eva, O.U., "Russkaya Pravoslavnaya Tserkov'posle Stalina, in *"Kruglyi stol": 50 let bez Stalina: nasledie stalinizma i ego vliyanie na istoriyu vtoroi poloviny XX veka*, 2003, March 4.

16   Vasil'eva, O.U. *Russkaya Pravoslavnaya Tserkov' v politike Sovetskogo gosudarstva v 1943–1948 godakh*, p. 108.

17   Ibid., p. 120.

18   Vasil'eva, O.U. *Russkaya Pravoslavnaya Tserkov' v politike Sovetskogo gosudarstva v 1943–1948 godakh*, p. 108

19   Ibid., p. 120.

20   Ibid., p. 207.

21   RGANI, F. 5, Op. 33, D. 91, L. 16.

22   Ibid., L. 90.

23   *Sovetskaya zhizn', 1945–53*, M.: ROSSPEN, 2003, p. 646.

24   RGANI, F. 5, Op. 33, D. 91, L. 17; Op. 34, D. 37, L. 4; D. 57, L. 23, 31.

25   The Soviet (Council) for the Affairs of the Russian Orthodox Church and Soviet (Council) for the Affairs of Religious Cults were merged in 1965 into a Soviet (Council) on Religious Affairs.

26   *Otechestvennaya istoriya*, 2002, no. 3, p. 47.

27   RGANI, F. 5, Op. 15, D. 429, L. 66–9.

28   RGANI, F. 5, Op. 16, D. 689, L. 114–15.

29   Ibid., L. 115.

30   Ibid.

31   RGANI, F. 5, Op. 16, D. 704, L. 197.

32   RGANI, F. 5, Op. 16, D. 690, L. 73.

33 RGANI, F. 5, Op. 34, D. 37, L. 74.
34 *Otechestvennaya istoriya*, 2002, no. 3, pp. 46–7.
35 GAVO, F. 1300, Op. 1, D. 954, L. 89.
36 RGANI, F. 5, Op. 34, D. 76, L. 8.
37 GAVO, F. 1300, Op. 1, D. 954, L. 89.
38 RGANI, F. 5, Op. 34, D. 76, L. 8.
39 RGANI, F. 5, Op. 34, D. 25, L. 59.
40 RGANI, F. 5, Op. 34, D. 57, L.133.
41 RGANI, F. 5, Op. 34, D. 57, L. 132.
42 RGANI, F. 5, Op. 33, D. 90, L. 54.
43 GAVO, F. 1300, Op. 1, D. 954, L. 90.
44 Ibid.
45 RGANI, F. 5, Op. 33, D. 90, L. 54.
46 RGANI, F. 5, Op. 34, D. 57, L. 133.
47 RGANI, F. 5, Op. 33, D. 90, L. 53
48 *Sovetskaya zhizn', 1945–1953*, p. 645.
49 RGANI, F. 5, Op. 33, D. 91, L. 66.
50 Ibid., L. 80.
51 RGANI, F. 5, Op. 34, D. 25, L. 70–1.
52 Ibid., L. 36.
53 Ibid., L. 36–7.
54 RGANI, F. 5, Op. 34, D. 120, L. 37.
55 RGANI, F. 5, Op. 55, D. 72, L. 109.
56 Ibid.
57 RGANI, F. 5, Op. 34, D. 120, L. 38–9.
58 Ibid.
59 RGANI, F. 5, Op. 33, D. 233, L. 40–1.
60 Koroleva, L.A., Korolev, A.A., *Gosudarstvo i religioznye ob'edineniya vo vtoroi polovine 1960–1980-kh godov (Penzenskaya oblast')*, M.: Penza, 2002, p. 67.
61 RGANI, F. 5, Op. 33, D. 233, L. 40–1.
62 Denisova, L.N., *Rural Russia: Economic, Social and Moral Crisis*, New York: Nova Science Publishers, 1995, p. 216.
63 RGANI, F. 5, Op. 33, D. 233, L. 8–9.
64 Borzenko, V.I., "Religiya v postkommunisticheskoi Rossii: novaya evangelizatsiya 100 let spustya," *Kuda idet Rossiya? Al'ternativy obchestvennogo razvitiya*, M., Edition 1, 1994, pp. 231–3.
65 *Vestnik statistiki*, 1992, no. 1, p. 53.
66 *Rossiya na rubezhe vekov*, M., 2000, pp. 143–5.
67 Aivazova, S., Kertman, G., *Zhenshchiny na randevu s rossiiskoi demokratiei*, M., 2001, p. 38.
68 *Rossiya na rubezhe vekov*, pp. 143–7, 162; Aivazova, S., Kertman, G., *Zhenshchiny na randevu s rossiiskoi demokratiei*, p. 38.
69 "Istoricheskaya pamyat' naselenie Rossii," *Otechestvennaya istoriya*, 2002, no. 3, pp. 198, 200.
70 *Rossiya na rubezhe vekov*, pp. 143–7, 162, 396–404.
71 *Rossiiskaya gazeta*, 1994, April 5.
72 *Rossiya v tsifrakh. 2003. Kratkii statisticheskii sbornik*, M., 2003, p. 49.
73 Televizionnaya peredacha rossiiskogo NTV.

16 Triple-burden lifestyle

1 Vyltsan, M.A., *Zhavershaiushchii etap sozdaniya kolkhoznogo stroya (1935–1937 gody)*, M., 1978, pp. 104–5; Sotsialisticheskoe zemledelie, 1937, February 8.

2 Manning, R.T., "Women in the Soviet Countryside on the Eve of World War II, 1935–1940" in Farnsworth, B., Viola, L. (Eds.), *Russian Peasant Women*, New York: Oxford University Press, 1992, pp. 214–16.
3 Zima, V.F., Golod v USSR 1946–47 godov: proiskhozhdenie i posledstviya, M., IRI RAN, 1996, p. 45.
4 *Istochnik*. 1997, no. 1, pp. 147–8.
5 Danilov, V.P., Kim, M.P., Tropkin, N.V. (Eds.), *Sovetskoe krest'yanstvo. Kratkii ocherk istorii (1917–1969)*, M.: Izdatel'stvo Politicheskoi literatury, 1979, p. 362.
6 Ibid., pp. 362, 378.
7 RGANI, F. 5, Op. 15, D. 408, L. 108.
8 VOANPI, F. 2522, Op. 17, D. 135, L. 5–6.
9 RGANI, F. 5, Op. 24, D. 530, L. 138; VOANPI, F. 2522, Op. 17, D. 135, L. 5–6.
10 GARF, F. P.-7928, Op. 2, D. 178, L. 41–2.
11 Beznin, M.A., Dimoni, T.M., "Sotsial'nyi protest kolkhoznogo krest'yanstva (vtoraya polovina 1940-kh-1960-e gody)," *Otechestvennaya istoriia*, 1999, no. 3, p. 86; Pyzhikov, A., "Sotsial'no psikhologicheskie aspekty obchestvennoi zhizni v gody 'ottepeli'," *Svobodnaya mysl'*, 2003, no. 6, p. 105.
12 VOANPI, F. 2522, Op. 11, D. 128, L. 119–24.
13 Sonin, M.Y., *Aktual'nye problemy ispol'zovaniya rabochei sily v USSR*, M., 1965, p. 195; Dodge, N.D., Feshbach, M., "The Role of Women in the Soviet Agriculture," p. 248.
14 *Sel'skoe khozyaistvo SSSR*, M., 1960, pp. 235–8, 240–3, 348, 350–1, 359.
15 Zaslavskaya, T., Ryvkina, R., "Sibirskaya derevnya: sotsial'nyi portret," *Sel'skaya nov'*, 1981, no. 2, p. 9.
16 *Voprosy ekonomiki*, 1966, no. 10, pp. 63, 65.
17 Savina, N.V., *Priusadebnoe khozyaistvo krest'yanstva Evropeiskogo Severa Possii v 1965-nachale 1980-kh godov. Dissertatsiya na soiskanie uchenoi stepeni kandidata istoricheskika nauk*, Vologda, 2000, pp. 70–1, 73.
18 Ibid., pp. 42–5; *Vestnik statistiki*, 1981, no. 2, p. 79.
19 Babaeva, L.V., Nel'son, L., "Delovaya aktivnost' zhench' in v novykh ekonomicheskikh strukturakh," *Sotsiologicheskie issledovania*, 1992, no. 5, pp. 107–8.
20 Koznova, I.E., "Gor'koe maslo reform: ob agrarnykh preobrazovaniyakh na Vologodskoi zemle," *Krest'yanovedenie. Teoriya. Istoriya. Sovremennost'. Uchenye zapiski*, M., 1999, p. 213.
21 Babaeva, L.V., Nel'son, L., *Delovaya aktivnost' zhench'in v novykh ekonomicheskikh strukturakh*, p. 109.
22 Staroverov, V.I., "Sovremennoe rossiiskoe krest'yanstvo i fermerstvo," *Krest'yanskoe khozyaistvo: istoriya i sovremennost'*, Vologda, 1992, Part 1, p. 169.
23 Sazonov, S.N., "Sovremennye tendentsii v razvitii fermerskogo dvizheniya v Rossii: illuzii i real'nost'," *Krest'yanovedenie. Teoriya. Istoriya. Sovremennost', Ezhegodnik*, M., 1996, p. 249.
24 Kostygova, T.M., Kosheleva, I.Y., op. cit., pp. 90–1.
25 Ibid.
26 Babaeva, L.V., Nel'son, L., "Delovaya aktivnost' zhench'in v novykh ekonomicheskikh strukturakh," p. 111.
27 *Zhenskoe dvizhenie v kontekste rossiiskoi istorii. Yubileinye chteniya, posvyachennye 90-letiiu Pervogo Vserossiiskogo zhenskogo s'ezda*, M., 1999, p. 54.
28 Nefedova, T., "Polnyi nazad ili polnyi vpered?" *Znanie—sila*, 1994, March, p. 33.
29 Zherebin, V.M., "Intensificatsiya domashnei ekonomiki—shag v proshloe ili v buduch'ee" in *Chelovek, sem'ia, dukhovnyi mir na puti k informatsinnomu obchestvu*, M., 1994, p. 62.

30 *Voprosy statistiki*, 2003, no. 3, p. 12; *Rossiia v tsifrakh*, 2003, p. 201.
31 *Voprosy statistiki*, 2003, no. 5, pp. 5–7; *Zhenshchiny i muzhchiny v Rossii. 2002*, Statisticheskii sbornik, M., 2002, p. 19.
32 *Chelovek i trud*, 2003, no. 10, p. 35.
33 Gordeev, A., "Pora podnimat'sya s kolen," *Argumenty i fakty*, 2003, no. 48 (1205), November.
34 Ibid.
35 Approximately 64.5 million USD; Nikolaev, M., Zamestitel' Predsedatelya Soveta Federatsii Federal'nogo Sobraniya RF, *Razvitie sela—strategicheskii prioritet // Chelovek i trud*, 2003, no. 10, p. 34.
36 *Rossiisko-britanskoe sotsiologicheskoe obsledovanie rossiiskikh dereven' v 1980–1990-e gody*. Proekt "Sotsial'naya struktura sovetskogo sela." Brak i sem'ya kak protsess vosproizvodstva pokolenii.
37 *Vestnik statistiki*, 1988, no. 1, p. 75.
38 *Prava zhenshchiin v Rossii: issledovanie real'noi praktiki ikh soblydeniya i massovogo soznaniya (po rezul'tatam issledovaniya v g. Rybinske Yaroslavskoi oblasti na osnove glubinnykh interv'yu*, M., 1998. p. 96.

17 Household chores

1 Gruzdeva, E.B., "Sovmech'enie zhench'inami professional'noi i semeinoi rolei: problemy i puti ikh resheniya," *Integratsiya zhench'in v protsess obch'estvennogo razvitiya*, M., 1994, pp. 337–9.
2 *Feminizm: perspektivy sotsial'nogo znaniya*, M., 1992
3 Dodge, N.D., Feshbach, M., "The Role of Women in the Soviet Agriculture," pp. 254–5; Yurkevich, N.G., "Rabota zhenshchiny na promyshlennom predpriyatii i stabil'nost' sem'i," *Materialy XII Mezhdunarodnogo seminara po issledovaniyu sem'i*, M., 1972, p. 25; Novikova, E.E., Yazykova, V.S., Yankova, Z.A., *Zhench'ina. Trud. Sem'ya (Sotsiologicheskii ocherk)*, M., 1978, pp. 60–1; Komarova, E.I., *Zhench'ina-rukovoditel'*, M., 1989, p. 61; Golod, S.I., *Sem'ya i brak: istoriko-sotsiologicheskii analiz*, SPt, 1988, p. 173.
4 Zdravomyslova, E., "Kollektivnaya biografiya sovremennykh rossiiskikh feministok (issledovanie na osnovanii 18 biograficheskikh interv'yu 1995–96 godov)," *Gendernoe izmerenie sotsial'noi i politicheskoi aktivnosti v perekhodnyi period*, SPt, 1996, p. 39.
5 Yastrebinskaya, G., Subbotina, O., "Istoriya sem'I Vorob'evykh. Vologodskaya oblast', Vologodskii raion, derevnya Utkino," *Otechestvennye zapiski*, 2002, no. 6, p. 522.
6 *Budzhet vremeni sel'skogo naseleniya*, M., 1979, pp. 125, 134.
7 "Istoriya sem'i Tolmachevykh iz sela Zyablovka Saratovskoi oblasti, *Golosa krest'yan: Sel'skaya Rossiya XX veka v krest'yanskikh memuarakh*, M., 1996, pp. 198–9.
8 GARF, F. 7928, Op. 2, D. 188, L. 210.
9 Krapivenskii, S.E., Dement'ev, S.M., Kramarev, V.F., *Sel'skokhozyaistvennyi kollektiv kak ob'ekt sotsial'nogo planirovaniya*, M., 1981, pp. 144–5.
10 Piskunov, V.T., *Priusadebnoe khozyastvo na sele*, M., 1983, p. 106.
11 *Sel'skaya nov'*, 1981, no. 12, p. 9.
12 Denisova, L.N., *Nevospolnimye poteri: Krizis kul'tury sela v 60–80-e gody*, M.: Nauka, 1995, p. 133.
13 Ryvkina, R.V., *Obraz zhizni sel'skogo naseleniya*, Novosibirsk, 1979, p. 198.
14 Anokhina, L.A., Shmeleva, M.N., *Kul'tura i byt kolkhoznikov Kalininskoi oblasti*, p. 213.
15 RGASPI, F. 591, Op. 1, D. 88, L.40.

208   *Notes*

16 Ibid., L. 39.
17 *Sel'skaya nov'*, 1989, no. 6, p. 7.
18 *Vestnik statistiki*, 1987, no. 5, p. 66.
19 RGANI, F. 5, Op. 16, D. 692, L. 16.
20 Denisova, L.N., *Ischezayuch'aya derevnya Rossii. Nechernozem'e v 1960–1980-e godi*, M.: Logos, 1996, p. 163
21 Dodge, N.D., Feshbach, M., "The Role of Women in the Soviet Agriculture," pp. 254–5.
22 RGANI, F. 5, Op. 15, D. 431, L. 36.
23 RGANI, F. 5, Op. 34, D. 57, L. 184.
24 *Plenum TsK KPSS, 24–26 marta 1965 goda*, Stenograficheskii otchet, M., 1965, p. 144.
25 Denisova, L.N., *Ischezayuch'aya derevnya Rossii*, p. 163.
26 *Rossiisko-britanskoe sotsiologicheskoe obsledovanie rossiiskikh dereven' 1990-h godov*, provedennoe G.A. Yastrebinskoi i O.G. Subbotinoi.
27 RGASPI, F. 591, Op. 1, D. 243, L.178; Krest'yanka, 1978, no. 9, p. 32.
28 *Krest'ianka*, 1980, no. 1, p. 17; no. 2, p. 24.
29 Peredacha rossiiskogo televideniya NTV. 2007 god.
30 RGASPI, F. 591, Op. 1, D. 90, L.8.
31 RGANI, F. 591, Op. 1, D. 58, L. 32.
32 Ibid., L. 67.
33 RGANI, F. 591, Op. 1, D. 243, L. 81.
34 RGANI, F. 591, Op. 1, D. 58, L. 26.
35 Ibid., L. 34–5.
36 RGANI, F. 591, Op. 1, D. 197, L. 103.
37 Anokhina, L.A., Shmeleva, M.N., *Kul'tura i byt kolkhoznikov Kalininskoi oblasti*, M., 1964, pp. 224–5.
38 Popov, A.A., "Izmeneniya v bytu sel'skogo naseleniya Severa Evropeiskoi chasti RSFSR v 60–70-e gody (na materialakh Komi derevni)," *Material'noe polozhenie, byt i kul'tura severnogo krestyanstva* (Sovetskii period), Vologda, 1992, p. 97.
39 *Narodnoe khoziastvo RSFSR v 1989 godu*, pp. 67, 213.
40 RGASPI, F. 591, Op. 1, D. 216, L. 124.
41 Narodnoe khozyastvo RSFSR v 1989 godu, p. 134.
42 *Vestnik statistiki*, 1991, no. 2, p. 60.
43 Narodnoe khozyastvo RSFSR v 1988 godu, pp. 102–3.
44 Narodnoe khozyastvo RSFSR v 1989 godu, pp. 198–9.
45 Ibid.
46 Dodge, N.D., Feshbach, M., "The Role of Women in the Soviet Agriculture," p. 254.
47 *Rossiisko-britanskoe sotsiologicheskoe obsledovanie rossiiskikh dereven' 1990-h godov*, provedennoe G.A. Yastrebinskoi i O.G. Subbotinoi.
48 Sovetskaya kul'tura, 1979, January 16; Sotsiologicheskie issledovaniya, 1982, no. 4, pp. 104–5.
49 *Sel'skie naselennye punkty RSFSR*, Stat, sb., M., 1970, Tom 1, pp. 9–10; *Vestnik statistiki*, 1992, no. 1, p. 10.
50 RGASPI, F. 591, Op. 1, D. 111, L. 93.
51 Denisova, L.N. *Nevospolnimye poteri: Krizis kul'tury sela v 60–80-e gody*, p. 95.
52 *Sotsiologicheskie issledovaniya*, 1980, no. 3, pp. 86–7; Alekseev, A.I., *Opyt konkretnogo sotsial'no-geograficheskogo obsluzhivaniya naseleniya*, p. 88; Zotova, O.I., Novikov, V.V., Shorokhova, E.V., *Osobennosti psikhologii krest'yanstva. Proshloe i nastoyashchee*, M., 1983, pp. 81–3.

53 Aivazova, S., Kertman, G., *Zhenshchiny na randevu s rossiiskoi demokratiei*, pp. 12–14; Lapin, N.I., *Kak chuvstvuyut sebya, k chemu stremyatsya grazhdane Rossii. Po rezul'tatam Vserossiiskikh issledovanii v ramkakh monitoringa "Nashi tsennosti i interesy segodnya": 1990, 1994, 1998, 2002 gody*. Kuda prishla Rossiya? Itogi sotsiental'noi transformatsii, 2003. Mezhdunarodnyi simpozium 16–18 yanvarya 2003 goda. Tezisy, M., 2002. M, MVShSEN, 2003.
54 Sotsiologicheskie issledovaniya, 1980, no. 3, p. 88; *Budzhet vremeni sel'skogo naseleniya*, M., 1979, p. 175; Zotova, O.I., Novikov, V.V., Shorokhova, E.V., *Osobennosti psikhologii krest'yanstva. Proshloe i nastoyashchee*, pp. 86–7.
55 RGANI, F. 5, Op. 35, D. 89, L. 94; RGANI, F. 5, Op. 37, D. 35, L. 112; RGANI, F. 5, Op. 37, D. 63, L. 77.

18 The special environment of the village life

1 *Rossiisko-britanskoe sotsiologicheskoe obsledovanie rossiiskikh dereven' v 1980–1990-e godi*. Proekt "Sotsial'naya struktura sovetskogo sela." Brak i sem'ya kak protsess vosproizvodstva pokolenii.
2 Zaslavskaya, T., Ryvkina, P., "Sibirskaya derevnya: sotsial'nyi portret," *Sel'skaya nov'*, 1981, no. 10, p. 7; no. 11, pp. 10–11.
3 *Velikii neznakomets. Krest'yane i fermery v sovremennom mire*, M., 1992, pp. 50–2.
4 *Sotsiologicheskie issledovaniya*, 1992, no. 5, p. 63.
5 RGANI, F. 5, Op. 15, D. 408, L. 75, 78.
6 Ibid., L. 108–9.
7 GAVO, F. 1300, Op. 1, D. 1327, L. 150 ob.-151.
8 Aivazova, S., Kertman, G., *Zhenshchiny na randevu s rossiiskoi demokratiei*, p. 27.
9 Pankratova, M.G. *Sel'skaya zhenchshina v USSR*, pp. 125–6; Pankratova, M., *Sel'skie zhiteli Rossii. Sud'by i sem'i v XX veke*, M., 1995, p. 51.
10 *British Journal of Psychiatry*, 1993, no. 148, pp. 23–6; Harpham T., Blue, I. (Eds.), *Urbanization and Mental Health in Developing Countries*, Athenaeum Press, 1995.
11 Muzhchiny i zhenshchiny Rossii, 2002, p. 55.
12 Mishel Gomel, K., *Natsii—za psikhicheskoe zdorov'e. V fokuse vnimaniya—zhench'ina (Vsemirnaya organizatsiya zdravookhraneniya)*, Kiev, 1998.

19 Protection of childhood and motherhood in the countryside

1 Kosarev, U.A., *Sotsial'noe strakhovanie v Rossii: na puti k reformam*, M., 1999, p. 19.
2 SU RSFSR, 1918, no. 89, p. 906.
3 This is closer to sanitarium than to sanatorium. SU RSFSR. 1917, no. 13, p. 188; 1919, no. 20, p. 238.
4 Bulleten' ofitsial'nykh rasporiazhenii i soobshchenii Narkomprosa, 1922, no. 75, p. 6.
5 SU RSFSR, 1924, no. 81, p. 813.
6 SZ SSSR, 1929, no. 71, p. 673.
7 Manning, R.T., "Women in the Soviet Countryside on the Eve of World War. 1935–1940" in *Russian Peasant Women*, New York: Oxford University Press, 1992, p. 216.
8 GARF, F. P-8131, Op. 23, D. 130, L. 1, 1 ob., 2.
9 GARF, F. P-8131, Op. 23, D. 99, L. 1, 1 ob.
10 GARF, F. P-8131, Op. 23, D. 130, L. 2.
11 Resheniia partii i pravitel'stva po khoziaistvennym voprosam (Sbornik dokumentov za 50 let), M., 1968, Tom 5, pp. 472–8.

12 Grigor'eva, N.S., Chubarova, T.V., *Gendernyi podkhod v zdravookhranenii*, M., 2001, pp. 19–20; *Upravlenie zdravookhraneniem*, 2003, no. 8, p. 19.
13 Sanatorium stays are usually for 21–4 days and include all meals free of charge; all meals are also provided free of charge in hospitals.
14 Bogdanov, P., K statistike i kazuistike boleznei polovykh organov krest'ianok Kirsanovskogo uezga, Tambov, 1989, p. 3.
15 Novikov, A., *Zapiski zemskogo nachal'nika*, SPt, 1899, p. 16.
16 Fedorov, V.A., "Mat' i detia v Russkoi derevne," *Vestnik MGU*, no. 4, 1994, p. 19.
17 *Okhrana materinstva i detstva v Rossii i Velikobritanii*, pp. 42–52.
18 *Moskovskii komsomolets*, 2006, September 28.
19 Malysheva, M.M., "Kommunisticheskaia utopiua v sud'bakh sel'skikh zhenshchin" in *Demografiia i sotsiologiia*, Edition 15, Gendernye aspekty sotsial'noi transformatsii, M., 1996, p. 222.
20 Narodnoe khoziaistvo RSFSR v 1980 godu, Statisticheskii ezhegodnik, M., 1981, p. 295.
21 Narodnoe khoziaistvo RSFSR v 1980 godu, p. 295; Narodnoe khoziaistvo RSFSR v 1988 godu, Statisticheskii ezhegodnik, M., 1989, pp. 217, 236; Narodnoe khoziaistvo RSFSR v 1989 godu, Statisticheskii ezhegodnik, M., 1990, p. 236.
22 *Rossiiskii statisticheskii ezhegodnik*, 2002, Statisticheskii sbornik, M., 2002, p. 245.
23 Shineleva, L.T., *Zhenshchina i obshchestvo. Deklaratsii i real'nost'*, M., 1990, p. 89.
24 *Vestnik statistiki*, 1991, no. 8, p. 58.
25 *Zhenshchiny v USSR. 1990, Statisticheskie materialy*, M., Finansi i statistika, 1990, pp. 8–9.
26 *Vestnik statistiki*, 1992, no. 1, p. 54.
27 *Vestnik statistiki*, 1991, no. 8, p. 58.
28 Ibid., no. 1, p. 63.
29 *Rossiiskii statisticheskii ezhegodnik*, 2002, p. 253; "Zhenshchiny i muzhchiny Rossi," 2002, *Statisticheskii sbornik*, M., 2002, pp. 33, 189–90.
30 Veselkova, I.N., "Sotsial'naia nestabil'nost' i sostoianie zdorov'ia naseleniia Rossii" in *Sotsial'nye koflikty: ekspertiza, prognozirovanie, tekhnologiia razresheniia*, M., 1999, p. 56; "Ob itogakh khoda reform i zadach po razvitiu zdravookhraneniia i meditsinskoi nauki v RF na 2000–2004 gody i na period do 2010 goda," Doklad ministra zdravookhraneniia RF U.A. Shevchenko, M., 2000, pp. 6–7; Okhrana materinstva i detstva v Rossii i Velikobritanii, p. 25; Romanov, K.S., "Kratkii analiz statistiki naseleniia i dinokhoziaistv v gendernom aspekte" in *Statisticheskoe issledovanie sotsial'no-demograficheskoi situatsii. Sbornik nauchnykh trudov*, M., 2002, p. 51.
31 *Sotsial'no-ekonomicheskie problemy agropromyshlennogo kompleksa Rossii*, M., 2000, pp. 15–16.
32 *Golosa krest'ian. Krest'iane Kubani pishut G.V. Iavlinskomu, Kuban'*, M., 1998–99, p. 56.
33 *Upravlenie zdravookhraneniem*, 2003, no. 8, pp. 18–25, 79.
34 *Rossiiskii statisticheskii ezhegodnik*, 2002, p. 125; Giriet, M.N., *Detoubiistvo. Sotsiologicheskoe i sravnitel'no-uridicheskoe issledovaniia*, M., 1911, p. 67.

20 Abortions

1 Geriet, M.N., *Detoubiistvo. Sotsiologicheskoe i sravnitel'no-iuridicheskoe issledovanie*, M., 1911, p. 67.
2 SU RSFSR, 1920, no. 90, Chapter 471; UK RSFSR, 1922, Chapter 146; ibid.; 1926. Chapter 140.

3 Goldman, W., "Women, Abortion and State. 1917–1936" in *Russia's Women. Accommodation, Resistance and Transformation*, Berkeley, 1991, pp. 243, 248, 249.

4 *Aborty v 1926 godu,* M., 1929, pp. 18, 36, 54; *Aborty v 1925 godu*, M., 1927, pp. 53–5, 64.

5 GARF, F. 396, Op. 3, D. 6, L. 17.

6 Mironov, B.N., Sotsial'naia istoriia Rossii perioda imperii (XVIII-nachalo XX vekov). Tom 1–2, SPt, Tom 1, 2000, p. 185.

7 Goldman, W., "Women, Abortion and State. 1917–1936," pp. 246, 250, 256–7, 261–2.

8 SZ SSSR, 1936, no. 34, Chapter 309.

9 *Gendernaia ekspertiza rossiiskogo zakonodatel'stva*, M., 2001, p. 188.

10 Manning, R.T., "Women in the Soviet Countryside on the Eve of World War II. 1935–1940," pp. 206, 226.

11 *Rabochii put' (Smolensk)*, 1936, May 30; Manning, R.T., "Women in the Soviet Countryside on the Eve of World War II. 1935–1940," p. 207.

12 *Pravda*, 1937, March 3; ibid., 1937, August 2; ibid., 1938, April 8; Manning, R.T., "Women in the Soviet Countryside on the Eve of World War II. 1935–1940," pp. 207, 227.

13 Vyltsan, M.A., *Zavershaushchii etap sozdaniia kolkhoznogo stroia (1935–1937 gody)*, M., 1978, p. 193.

14 Manning, R.T., "Women in the Soviet Countryside on the Eve of World War II. 1935–1940," p. 208.

15 Shargorodskii, M.D., *Prestuplenie protiv zhizni i zdorov'ia*, M., 1948, pp. 90, 421.

16 *Sotsialisticheskoe zemledelie,* 1936, June 5; *Pravda*, 1938, December 14; Manning, R.T., "Women in the Soviet Countryside on the Eve of World War II. 1935–1940," p. 207.

17 *Pravda*, 1936, September 15.

18 GARF, F. 5446, Op. 29, D. 361, L.1–6.

19 *Sotsialisticheskoe zemledelie*, 1936, June 1, 5.

20 *Sotsialisticheskoe zemledelie*, 1936, June 1, 5, 6, 10; *Pravda*, 1936, May 28–31, June 1–6, July 8, 11, 12; *Krest'ianskaia gazeta*, 1936, June 8; also Manning, R.T., "Women in the Soviet Countryside on the Eve of World War II. 1935–1940," p. 208.

21 Manning, R.T., "Zhenshchiny sovetskoi derevni nakanune Vtoroi mirovoi voiny. 1935–1940 gody" in *Otechestvennaia istiriia*, 2001, no. 5, p. 89.

22 *Sotsialisticheskoe zemledelie*, 1936, June 1, 5; Manning, R.T., "Women in the Soviet Countryside on the Eve of World War II. 1935–1940," pp. 208–9.

23 *Trud*, 1936, April 27.

24 *Sovetskaia zhizn'. 1945–1953*, M., Rosspen, 2003, p. 681.

25 "Zhenshchina, kotoraia razreshila aborty" in *Moskovskii komsomolets*, 2004, August 24.

26 VOANPI, F. 2522, Op. 17, D. 98, L. 32–3.

27 *Sovetskaia zhizn'. 1945–1953*, M., Rosspen, 2003, p. 685.

28 Ibid., pp. 682–3.

29 Dagel', P.S., "Usloviai ustanovleniia ugolovnoi otvetstvennosti" in *Pravovedenie*, 1975, no. 4, p. 70.

30 Lunev, V.V., *Prestupnost' XX veka. Mirovye, regional'nye i rossiiskie tendentsii, Mirovoi kriminalisticheskii analiz*, M., 1997, p. 206.

31 *Vedomosti Verkhovnogo Soveta SSSR*, 1954, no. 15, Chapter 334.

32 Bridger, S., "Zhenshchina v sovremennom sovetskom obshchestve" in *Vestnik RAN*, 1992, no. 2, p. 85.

33 *Rossiiskii statisticheskii ezhegodnik*, 2002, p. 246.

34 *Vestnik statistiki*, 1991, no. 8, p. 61.

35 *We*, 1998, no. 5 (21), p. 31; Serov, V.N., "Temnaia storona seksa," *Vek*, 2001, no. 12.
36 *Zhenskii vopros. 37 osnovnykh tem o zdorov'e*, SPt, 2002, p. 77.
37 *Rossiiskii statisticheskii ezhegodnik*, 2002, p. 212.
38 *Moskovskii komsomolets*, 2006, May 12.
39 *Argumenty i fakty*, 2004, no. 30.
40 Shlidman, Sh., Zvidrinysh, P., Izuchenie rozhdarmosti (po materialam spetsial'nogo issledovaniia v Latviiskoi SSR), M., 1973, p. 57; *Argumenty i fakty*, 2004, no. 40, p. 20.
41 *Zdorovie zhenschiny Rossii*, p. 9.
42 "Tezisy RUSO. O sovremennom demograficheskom bedstvii v Rossii i merakh po ego preodoleniyu" in *Burevestnik*, 2001, May–June, no. 4–5, p. 8.
43 *Rodnaia gazeta*, 2003, November 14.
44 Il'in, V.A., Shabunova, A.A., *Problemy zdorovogo detstva: regional'nyi aspect*, Vologda, 2000.
45 We have already discussed some problems related to children born out of wedlock in Chapter 9, especially their legal status and the interwar period.
46 Iakovlev, G.V., *Okhrana prav nezamuzhnikh zhenshchin*, Minsk, 1979, p. 7.
47 Prava zhenshchin v Rossii: issledovanie real'noi praktiki ikh soblyudeniia i massovogo soznaniia (po rezul'tatam issledovaniia v g. Rybinske Iaroslavskoi oblasti na osnove glubinnykh interv'yu), M., 1998, p. 97.
48 Antipova, M.L., "Vnebrachnye deti v Rossii" in *Statisticheskoe issledovanie sotsial'no-demograficheskoi situatsii. Sbornik nauchnykh trudov*, M., 2002, pp. 5–6; *Rossiiskii statisticheskii ezhegodnik*, 2002, p. 125. It needs to be noted that even though this number seems high, it nonetheless lags significantly behind many Western European countries and the US (e.g. as of 2007, 40 percent of all children born in the US are to unmarried mothers; 66 percent for Iceland; 55 percent for Sweden; 50 percent for France, and 44 percent for the UK).
49 Mikheeva, A.R., "Fenomen sozhitel'stva v sibirskoi derevne: novaia forma sem'i ili prodolzhenie traditsii?" in *Sotsiologicheskie aspekty perekhoda k rynochnoi ekonomike (Materialy k XIII Vsemirnomu sotsiologicheskomu kongressu). 4.1. Problemy sotsial'noi adaptatsii k izmeniayushchimsia usloviiam zhizni*, Novosibirsk, 1994, p. 142.
50 *Naselenie Rossii na rubezhe XX–XXI vekov. Problemy i perspektivy*, M., 2002, p. 83.
51 Ibid.; Antipova, M.L., "Vnebrachnye deti v Rossii" in *Statisticheskoe issledovanie sotsial'no-demograficheskoi situatsii. Sbornik nauchnykh trudov*, M., 2002, pp. 7–8.

Index

Fedorov 135
female alcoholism 117–20
finances, family 163
flax production 13–14
folk songs and rhymes 17, 20, 87, 113, 133
freedom of movement 63, 124
freedom of religion 139, 141
fumes 24, 38

gardening plots 2, 22, 48, 57, 137, 143, 145, 149, 184
Garmash, Daria 28
gender equality 16, 24, 67–9, 80–2, 88, 104, 105, 151, 167; and work 10, 13, 17, 20, 41, 110, 146–7, 155
gender imbalance 42, 56, 61–2, 89, 135, 153
generational divide 20, 133, 136
geographical focus of the book 2
Globa, M. 17
Godunova 31
Gorbachev era 3, 82, 118, 141
Gorkii region 138, 157, 166
The Great Terror 55, 124, 132
Gribanova, Z.A. 29
Grozny region 57
guberniia 112
GULAG 124–5
Gureeva, M.A. 29
Gzhel 185

Halimi, Gisèle 108
happiness criteria 91, 92
hard currency prostitutes 129–30
Harvest of Sorrow famine 123
Heroes of Labor 29
honors and awards 15, 25–6, 29, 34, 75
hours worked 9, 10, 18, 21, 24, 27, 32, 37, 38, 49–51, 60, 81, 146–7, 155
household appliances 158–9
household chores 2, 9, 36, 143–60
housewives 69
housing 45, 47, 57, 153–4

illegitimate children 3, 68, 75–7, 78, 128, 180, 183
immigration 130

industrialization 7, 22
infanticide 181
inflation 3, 83
Institute of Obstetrics and Gynecology 170
Institute of Pediatrics 170
Institute for Social Studies of the Russian Academy of Sciences 162
internet 53
Irkutsk region 49
Isliukov, S.M. 78
Istobniki 49
Ivanovo-Voznesenks region 73, 177
Ivanovo region 125
Izvestiia 79

Judiciary Committee of the Soviet of Ministers 78

K., Olga 97
Kalinin region 36, 41, 57, 63, 138, 139, 141, 144, 185
Kaliningrad region 57
Kalinova, Olga 85
Kaluga region 57, 154
Kaminskii, G.N. 179
Kamyshinsk district 58
Karelia 57, 58
Karpovskaia 138
Kazakova, M.F. 125
Kemerovo region 16
Khrushchev era 3, 42, 59, 79, 140
Kichmengsko-Gorodetskii district 21
Kiev region 43
Kirov region 19, 135, 157
Kislovo 156
Komsomol 14, 19, 38, 47, 60, 124, 132, 139
Komsomol'skaia Pravda 62
Kornilova, A. 42
Koshuniaeva, G.A. 32
Kostroma region 12, 58, 62, 185
Kovardak, P. 17
Krasnodar region 27
Krasnyi Putilovets 144
Krestianka 20, 21, 29
Kronstadt 168